Praise for *Younger*

"Is our health destiny a manifestation of nature or nurture? Are we the product of our genetic inheritance or our lifestyle decisions? In *Younger*, Dr. Sara Gottfried reveals that what most influences whether we are healthy versus ill—and the grace by which we age—is the dance between our lifestyle choices and how they influence gene expression. *Younger* fully delivers on the promise to provide the tools to actually change your genetic expression to regain health, stave off disease, and live with exuberance."

—David Perlmutter, M.D., *New York Times* bestselling author of
Grain Brain and *The Grain Brain Whole Life Plan*

"There is a way to grow more beautiful and feel stronger with each passing year. It isn't about diets or drugs or any magic formula; it is about the power of lifestyle medicine combined with the new science of aging. With Dr. Sara Gottfried's brilliant book *Younger,* we finally have the gold standard for how to combat aging—the first book of its kind actually rooted in science."

—Alejandro Junger, M.D., *New York Times* bestselling author of *Clean*

"Sarah Gottfried speaks openly to the secret question of a generation: 'Now that at last we have things figured out, how do we stay on the planet long enough to live what we finally know?' Her guidance on how to increase not only our life span but also our health span is an invaluable resource to those of us who feel we are spiritually youthing as we physically age."

—Marianne Williamson, *New York Times* bestselling author

"*Younger* is the breakthrough book we've been waiting for on DNA, epigenetics, and aging. It is a stunning achievement by one of our wisest and most thoughtful women physicians. Don't wait for a scary diagnosis—start the *Younger* protocol without delay."

—Mark Hyman, M.D., *New York Times* bestselling author and director
of the Cleveland Clinic Center for Functional Medicine

"Mind blown. Prepare to completely shift your paradigm around aging. Dr. Sara makes the latest science accessible to show you how to de-age your body and lengthen your health span. This book is a stunning, epic achievement."

—JJ Virgin, *New York Times* bestselling author of *The Virgin Diet*

"If you'd like to be smarter and more well-informed than your doctor about healthy aging and your own personal genetic and environmental risk factors, run out right now and grab your copy of Sara Gottfried's *Younger*. This remarkable, up-to-date, and scientifically rigorous book gives you all the tools you need for vitality, flexibility, strength, and happiness at any age. I plan to use *Younger* as my own personal health reference book."

—Rachel Carlton Abrams, M.D., author of *BodyWise: Discovering Your Body's Intelligence for Lifelong Health and Healing*

"*Younger* by Dr. Sara Gottfried is a must read for any adult who is hoping to age more gracefully. Whether you are just starting your health journey or you are an advanced enthusiast, Dr. Sara will take care of you. She'll guide you through the basic steps and offer some 'advanced challenges' for those who find themselves inspired. I loved the specific case studies, gene-healing recipes, and Dr. Sara's inspirational tone. Yes, your genes are the cards you're dealt. Yet, with Dr. Sara's help, you'll learn that you can choose how to play those cards."

—Dr. Alan Christianson, *New York Times* bestselling author of *The Adrenal Reset Diet*

YOUNGER

Also by Sara Gottfried, M.D.

The Hormone Reset Diet

The Hormone Cure

YOUNGER

A Breakthrough Program to
Reset Your Genes, Reverse
Aging, and Turn Back
the Clock 10 Years

SARA GOTTFRIED, M.D.

HarperOne
An Imprint of HarperCollinsPublishers

HarperOne

YOUNGER. Copyright © 2017 by Sara Gottfried. All rights reserved. Printed in the United States of America. No part of this book may be used or reproduced in any manner whatsoever without written permission except in the case of brief quotations embodied in critical articles and reviews. For information, address HarperCollins Publishers, 195 Broadway, New York, NY 10007.

HarperCollins books may be purchased for educational, business, or sales promotional use. For information, please email the Special Markets Department at SPsales@harpercollins.com.

FIRST HARPERCOLLINS PAPERBACK EDITION PUBLISHED IN 2018

Art by Kevin Plottner and Sara Gottfried
Designed by SBI Book Arts, LLC

Library of Congress Cataloging-in-Publication Data is available upon request.

ISBN 978-0-06-231628-8

18 19 20 21 22 LSC(H) 10 9 8 7 6 5 4 3 2 1

*Dedicated to my beloved patients
and community.*

*Thank you for teaching me the mysteries
of the human body.*

CONTENTS

Women, Aging, and Genetics

The laws of genetics apply even if you refuse to learn them.

—Alison Plowden

'm no supermodel. In fact, obesity, hair loss, anxiety, and Alzheimer's disease run in my family—not a pretty genetic picture for middle age and later life. My mother ate sparingly while pregnant with me, as was the fashion in 1967, an era of Twiggy and miniskirts. Mom's diet turned on my famine genes while my chromosomes were being knit in her womb, meaning that I've had a lifelong struggle with blood-sugar problems and rapid weight gain (a lot more on these topics later). I've grown up idolizing actresses such as Katharine Hepburn, Sigourney Weaver, Diane Keaton, and Julia Roberts. They were slim and tall, but I was pudgy and short.

Now, when I start to wonder why it's so freaking hard to stay mentally and physically fit at fifty, I remind myself that my genes program me to be a two-hundred-pound anxious diabetic with thinning hair. All things considered, maybe I'm not doing so badly.

Think about Angelina Jolie, Jennifer Lopez, Julianne Moore, Gisele Bündchen, and Helen Mirren. It's easy to believe they won the genetic lottery. Perhaps they hail from a long line of superwomen with flawless skin, flat bellies, perfectly balanced hormones, and fast metabolisms.

It's their job to look amazing, and they are extremely motivated to look good as long as possible as they age. Their taut abdominals and gravity-defying posteriors grace billboards, Victoria's Secret catalogs, and *Sports Illustrated* covers. They have proportionally similar metrics: Gisele Bündchen, the world's highest-paid model, is five feet eleven and weighs 126 pounds, and her bust-waist-hip measurements (in inches) are 35–23–35. Angelina Jolie is five feet eight, weighs 128 pounds, and measures 36–27–36. Their paychecks and magazine gigs depend on their enviable measurements. Even in her sixties, at five feet four and with 37–27–38 measurements, Helen Mirren rocked a coral bikini on a beach in Italy, looking better than me and most of my girlfriends.

That's great for those women, but the rest of us flounder. I don't know about you, but sometimes I feel like I was born to wrestle with my weight, skin, energy, and sex drive. In college, my weight ballooned. In medical school, my skin broke out and my adrenals broke down because of stress. I craved sugar and carbs and rarely ate vegetables. I drank gallons of coffee, hardly slept for a decade, and bought fat jeans. Then I had two kids. Need I go on?

Maybe you've been told that your muffin top or memory problems aren't your fault; they are simply programmed by your genes. It doesn't seem fair. When I was in my forties, my battle only seemed to get harder as I navigated the challenges of crazy work hours, perimenopause, grief, breast lumps, aging parents, tight clothes, travel, and stress. Eventually, I learned that there's a spiritual lesson in my battle with age and that my mess is my message.

The female body is magnificent, but it doesn't come with a lifetime warranty or an owner's manual. You're the result of millions of years of evolution, but many of the adaptations that helped your ancestors survive are now making you fat and wrinkly and are no longer needed. But your genetic code—the DNA sequence that is the biochemical basis of heredity in all living organisms—is only a small part of the story. Your DNA is a unique, one-of-a-kind blueprint that is specific to you. Even if you haven't been dealt platinum genes, you can still look great and age more slowly.

The fact is that scientists have found new ways for us to take control of our genes. For example, the naughty aging genes usually associated with fat and wrinkles can be altered with diet, exercise, and other lifestyle choices. Simply put, by turning your good genes on and your bad genes off, you can actually prevent aging—no matter how old you are.

Gisele's measurements are unattainable for the average American woman—who is five four and 164 pounds with a thirty-eight-inch waist—but even if you have fewer of the good genes and more of the bad genes, you can still lose weight, improve your skin, and change how your DNA controls your body and mind. You don't even need a large staff of trainers and chefs to hold you to your exercise regimen and diet; you can appear to have lucky genes whether or not you actually have them.

The truth is that around 90 percent of the signs of aging and disease are caused by lifestyle (and the environment created by your lifestyle), not genes.[1] The *neighborhood* of your body—how you live and the world you create, internally and externally—is more important than your DNA when it comes to how you look and feel now and for the next twenty-five to fifty years. So let's clean up your neighborhood.

Scientific Breakthroughs Make Staying Young Possible

I am a physician who trained at Harvard and MIT, but I was never taught the secrets to staying young. I didn't learn about them in medical school because many of them weren't yet discovered. It took a confluence of factors to create a new protocol for aging slowly. It took the Human Genome Project, which wasn't completed until 2003. It took affordable genetic testing, testing that up until five years ago cost about ten thousand dollars and now costs about two hundred dollars. It took bigger and better computers that could handle the volume of data that genomes provide—data sets that are so large and complex that novel data-processing applications had to be invented. It took me personally testing myself and, through trial and error, finding the genetic switches that control metabolism, weight, disease, and aging. Then it took refining my protocol for thousands of patients and women who work with me online before I learned the best, most scientifically proven ways to reprogram genes with specific lifestyle and mind-set changes.

In the process, I discovered what helps people not just *look* young but also *feel* young, and, even more exciting, I learned how DNA plays a part in overall aging and what we can do to alter the way our DNA is expressed. Who wouldn't want to influence her genes for the better?

Some women have asked me how this book is different from my previous books, *The Hormone Cure* and *The Hormone Reset Diet*. The first two books focus on your hormones, but this book will show you how to overcome and transform your genetic history and tendencies, particularly when it comes to aging. Feel destined for cellulite, saddlebags, and belly fat? Nothing seems to help your aging skin or declining libido or flagging energy? Does your family have a long history of Alzheimer's, cancer, or heart disease? This book is for you. Let's expand not only your life span but also your *healthspan*—the period during which you are able to thrive, free from disease and in hormonal harmony. Whether you are thirty-five or sixty-five, this protocol will help you prevent signs of aging and feel healthier and stronger than ever before.

The strategy of the Younger protocol is to interpret the warning signs of age in your body—the worsening vision, the thinner skin, the weaker lungs, the faulty memory—and turn them around. My goal is not to pick off one disease at a time (such as Alzheimer's, diabetes, or age-related cancers) but to delay or prevent all of these conditions, since they have a similar root cause: aging in any form. This means that by delaying one condition, you delay them all. This is the basis of *functional medicine*, the emerging system of medicine that engages the whole person, not just an isolated set of symptoms, and works from the inside out to address the root cause of disease and accelerated aging.

Plump Cheeks

I became interested in my genetic contract—the rules of my DNA and how they are expressed in my body—when I was thirty-nine. Something happened that I didn't expect: my cells started to betray me.

Let me explain. I was at a decent weight with a body mass index (BMI) of 25, right on the border between normal and overweight. I never thought of myself as middle-aged, but there I was, facing down forty, the official threshold. (*Middle age* is defined as age forty to sixty-five.) I heard from friends and family that I needed to get to my ideal body weight before forty because my metabolism would slow down precipitously then and future weight loss would come not from my belly, but from my face.

Apparently, in the physics of aging, volume in the face equals youthful vigor. Dermatologists even have a term for it: the *triangle of youth*. If you draw a line across the cheeks from ear to ear and then close the triangle by drawing a line from each ear to the chin, the widest part of the face is at the cheeks. But as you age, thanks to gravity, cheeks deflate, and fat moves south. Your body makes less collagen, and the collagen that it does make is less elastic, so your skin is not as thick and firm as before. Your bones thin, so the cheekbones shrink. Excess skin moves to the jaw, and now the widest part of your face is at the jawline, and the triangle of youth flips upside down.

Was this true? I decided to separate fact from fiction by applying my medical knowledge to my own aging body, just as I had done previously with hormones.

During this investigative period, I learned many surprising truths about aging. Much to my dismay, I discovered that fat loss does indeed occur in a woman's face, as opposed to her belly, after a certain age because collagen no longer undergirds the architecture of the facial skin and bones. Even so, I also learned that modulating estrogen levels with targeted lifestyle changes can slow down the loss of collagen. For instance, you can drink a collagen latte (see chapter 5) to boost production of collagen type III. Despite what you may think, not everything is inevitable when it comes to aging. I can promise you that we do have considerable control over the process.

Aging Accelerates at Forty

Let's look at what's actually happening in your body. By the time you reach middle age, there has been an unseen, predetermined, twenty-five-year process of cellular decline. (Don't freak out—I will show you how to circumvent this problem, no matter how close or far you are from forty.) Cellular decline progresses insidiously, unobserved by most people, perhaps including you and your well-intentioned doctor. You may notice it as muscle tightness, an emerging paunch, lingering hangovers, or difficulty reading labels, or you may recognize it by the fact that staying in shape seems to require ten times the effort. Your endocrine glands, from your ovaries to your thyroid, start to sputter and gasp in their hormone production. Then muscle mass declines and gets replaced by fat, and suddenly you realize—like I did on a recent fitness spree—that the activity of jumping is no longer an option. You start waking up at four in the morning for no good reason. Words you've used for decades evade you.

Unlike a fine Bordeaux, your body does not get better with age. Before you pour yourself another glass of wine to lament the facts of middle age, permit me to share some good news with you. Due to recent scientific breakthroughs, middle age now offers you a profound opportunity to re-program your genes and your body. I urge you to take this seriously before decay, or what we can refer to as *accelerated aging*, sets in, leading to not just minor annoyances like hair loss but also alarming diseases such as Alzheimer's and breast cancer. In fact, the Centers for Disease Control and Prevention reported that in 2015, for the first time in several years, longevity declined, due to an uptick in heart disease, diabetes, stroke, and Alzheimer's.[2] If those diagnoses seem abstract and irrelevant to you now, consider that by the year 2030, 20 percent of the population will be sixty-five or older (compared with 13 percent in 2010).[3] New cases of Alzheimer's will rise by 35 percent,[4] while new cases of breast cancer are expected to rise by 50 percent.[5] You don't want to be included in those statistics.

Using my medical education and practice as well as my own very personal struggle as a woman in a middle-aged body, I have developed a seven-week program called the Younger protocol to change the course of your aging body and grow your healthspan.

Five Aging Factors Gone Wrong

After forty, you begin to feel the effects of getting older. You can't indulge in French fries, sugary cocktails, and ice cream—or if you do, you can't get away with it. Gray hairs show up. When you're on your feet most of the day, leg veins no longer snap shut, and an unmistakable bulge of fluid collects at your ankles. You can't read your smartwatch without glasses (happened to me last week!). Your hormones are suddenly out of whack, and you find yourself sad, moody, tired, or chubby for unclear reasons. Your back goes out when you travel. You're not as stress resilient, and when you get a lousy night of sleep, you don't bounce back as easily as you once did. Why? Five key factors make aging more pronounced after forty, leading to *inflammaging*—the unfortunate hybrid of increasing inflammation, stiffness, and accelerated aging. Keep in mind it's not your age that's the enemy, it's the loss of function, and these are the culprits:

1. The Muscle Factor. Your metabolism slows down with age, which means you accumulate more fat and lose muscle. Think of aging as beginning in your muscles. The decline may not be noticeable at first, but on average, you lose five pounds of muscle every decade, so you definitely start to observe the change over the course of middle age. On the cellular level, your mitochondria become tired, a process known as mitochondrial dysfunction, which may make you feel more fatigued during and after exercise or cause muscle pain. Your mitochondria are the tiny powerhouses inside your cells that turn food and oxygen into energy. You have a thousand or two mitochondria inside most of your cells, and if they're gunked up with debris and damage, you will feel tired and achy. Causes range from eating empty calories such as sugar, flour, and overly processed foods to exposures to toxins. In summary, if left alone or ignored, your muscles usually get more doughy as they're replaced with fat, and you're not as strong as you used to be. The key is to focus on preserving and building your muscle mass as you age past forty.

2. The Brain Factor. Your neurons (nerve cells) lose speed and flexibility. Alcohol makes you foggier than it did, and you lose sleep. Connections between neurons, called synapses, are not what they used to be, so finding words may become an issue. The balance shifts toward more forgetting and less remembering. Part of the problem is that your brain gathers rust like an old truck left in the rain; free radicals induce damage to cells, DNA, and proteins in a process called oxidative stress if you don't have antioxidant countermeasures in place (like vitamins A, C, and E). Research indicates that if you're female and around forty-three or older (that is, in perimenopause), your brain becomes resistant to the lubricating and mood-lifting benefits of estrogen. Gluten, found in wheat and flour products, may make the problem worse. Your hippocampus—the part of your brain involved in memory creation and emotional control—may shrink, especially if you're stressed. As if that weren't bad enough, excess stress kills brain cells by increasing production of beta-amyloid, which then forms disruptive plaques that harm synapses further, putting the brain at risk for Alzheimer's disease. The key is to focus on keeping your brain regenerating and malleable (or "plastic") as you get older.

3. The Hormone Factor. Your hormones change for the worse. With age, both men and women make less testosterone, leading to more fat deposits at the breasts, hips, and buttocks. Women produce less estrogen, which normally protects the hair follicles and skin. Lower estrogen-to-testosterone ratios may trigger hair loss and heart disease. Unfortunately, your thyroid gland slows down and, along with it, your metabolism, so the bathroom scale climbs a few pounds per year (or even per month). You get cold more easily. Your thyroid may become lumpy or attack itself. Your cells become increasingly insensitive to the hormone insulin, which leads to rising blood sugar in the morning. (After you hit age fifty, blood-sugar levels rise approximately 10 mg/dL every decade.) As a result of higher blood sugar, you may feel foggier and experience stronger

cravings for carbs, then notice more skin wrinkling along with an older-looking facial appearance.[6] Older adults are less able to maintain sleep, leading to chronic sleep deprivation, which results in more wear-and-tear hormones (e.g., cortisol) and fewer growth-and-repair hormones (e.g., growth hormone). More cortisol and less growth hormone translate into even more skin wrinkling, facial aging, and higher morbidity and mortality.[7] Lower levels of estrogen and testosterone may weaken your bones *and* your sex drive. The key point is that the right food, sleep, exercise, and support for detoxification can reverse many hormone problems associated with aging.

4. The Gut Factor. Of course, there's overlap between these various factors. About 70 percent of your immune system lies beneath your gut lining, so it's the place where your immune system can get overstimulated, leading to excess inflammation and even autoimmune conditions. Your gastrointestinal tract contains three to five pounds of microbes, mostly bacteria and a small amount of yeast, that exist in your mucosa from your mouth to your anus. The DNA from your microbes outnumber your human DNA a hundred to one and are collectively known as your *microbiome*. Several studies show that your microbiome may affect your hormones, including estrogen and testosterone. Imbalanced microbes and their DNA may cause you to make more enzymes such as beta-glucuronidase, which raises certain bad estrogens and lowers your protective estrogens. Further, excess stress raises corticotropin-releasing factor, which pokes holes in your gut, leading to food intolerances, more stress, and lower vagal tone, an indicator that your nervous system is out of whack. Finally, high stress can make you absorb nutrients poorly, especially B vitamins; it's as if your body requires a full parking lot to function well and age slowly, and the missing B vitamins are empty parking spaces, waiting to get filled. But don't get lost in the details; just know that your gut can accelerate or decelerate your clock.

5. The Toxic Fat Factor. When you're trying to preserve your youth and health, toxins from the environment accumulate in your fat. Scientists call them *gerontogens*. They are similar to how carcinogens increase your risk of cancer, and they can work against you and cause premature aging. Pollution, cigarette smoke, heavy metals, UV rays, chemotherapy, contaminated drinking water, preservatives, and pesticides can all conspire against you. Take chemo for breast cancer as an example—it may add fifteen years to your chronological age, so that you die earlier but without cancer. Additionally, fat deposited in your belly is biochemically different than fat elsewhere; it makes an inflammatory brew of bad chemicals that causes you to age faster than someone who has only minimal visceral fat. While exposure to certain poisons are inevitable, we can attack the genetic flaws that cause you to accumulate them.

**FIVE FACTORS THAT
STEAL YOUR YOUTH**

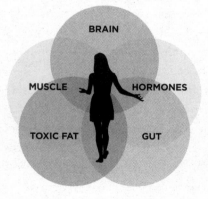

The net result of these five factors is a vicious cycle of more inflammation, an overactive immune system that's ready to attack normal tissues, and faster aging. In the following chapters, you'll learn how to disarm, prevent, and reverse these five factors and change the expression of genes that influence them. If you are tired of the sinking feeling that you are getting older, slower, and fatter by the day, turn the page and discover how you can unlock your genes and live longer, stronger, and better than ever.

CHAPTER 1

Unlock Your Genes

I grew up, and old.

—Alice Munro, "Some Women"

C rucial truths are locked in your genes. When unlocked, they will forever alter the way you think about how your body ages. I started at age thirty-nine to upgrade my lifestyle and throughout my forties found benefits I never thought possible. More energy—I'll take it. Less stress—that was a no-brainer. Glowing, thicker skin—*yes, please!*

Allow me to be perfectly clear: I'm not interested in the quackery or unproven potions that are rumored to help you stave off mortality. I'm not going to suggest you start injecting yourself with age-defying hormones. Instead, I focus on the rigorous methods of smart, unbiased researchers who are optimistic about which natural efforts can result in extended vitality as well as on the cultural insights from long-lived populations around the world. I'm especially fascinated by the practical experience of my own patients, and I've observed which habits are the easiest to adopt and, therefore, the most successful.

Yes, aging is inevitable. But you can delay accelerated and unnecessary aging so that it's a far slower and richer journey than the steady march of

Meet Heather: No Fillers Required

Heather was a forty-five-year-old teacher who came to see me because of early bone loss and ten stubborn pounds she couldn't lose. She asked me to fix her wrinkles too, because as a single woman, she had difficulty in the online dating world competing against women in their thirties with glowing skin. Encouraged by some friends, she went to an eye doctor who no longer practiced ophthalmology but instead injected fillers into aging women's faces and necks. She came away a bit horrified by how beat up she looked for days, not to mention by the tremendous out-of-pocket expense of five hundred to a thousand dollars per area.

"Is this what it has come to, Dr. Sara? Is there an alternative to fillers?" she asked.

I made a few small changes to Heather's lifestyle. I asked her to eat more marine fat—from cold-water, wild-caught fish and seafood—and to floss twice per day. She drank bone broth during her hours in the classroom and used it as a base for soups. We added vitamin D and DHEA to her daily regimen. She scheduled two yoga classes per week and frequent brisk walks. She returned after eight weeks, thrilled with the results. The weight fell off. Her skin was smoother. Heather radiated confidence. She looked profoundly younger, no fillers required.

decline that too many women experience. I've discovered the exceptions to the rules of aging so that you can disobey them too.

For many years, I've been using functional medicine with clients and on myself. It is an extremely therapeutic approach that creates a healing partnership between patient and practitioner. In my professional opinion, this has been the missing piece in modern medicine. Disease results from too much or too little of anything in life. Functional medicine looks at the interactions among genetic, environmental, and lifestyle factors that can influence long-term health and disease. Having been educated in both paradigms, my opinion is that traditional medicine is essential when you have a broken bone or pneumonia, but functional medicine may be a better approach when it comes to preventing and reversing chronic disease. We want to find out if functional medicine works better or not studied against conventional medicine. The Cleveland Clinic is now performing clinical trials comparing head-to-head the standard medical treatment versus functional medical care for asthma, inflammatory bowel disease, migraines, and type 2 diabetes—watch for results in the years to come.[1]

We've Got It Backward

Here's a sobering statistic I learned from Bill Gifford, science writer and author of *Spring Chicken: Stay Young Forever (or Die Trying)*: consumers spend more on plastic surgery than the government spends on aging research.[2] I'm not judging—far from it. Even though I am Dr. Natural, sometimes I catch myself looking for a long while at the noninvasive or surgical cosmetic solutions advertised on billboards and TV. What about you?

Two of the most common plastic surgeries, eyelid surgery and facelifts, are growing in popularity.[3] Why are antiaging procedures on the rise? What motivates people to have plastic surgery as they age?[4] Scientific studies show that the people most likely to go under the knife are wealthy females with low self-esteem, low life satisfaction, low self-rated attractiveness, and few religious beliefs. These women are also heavy television watchers (and perhaps see beautiful bodies on TV that they want to emulate). The common motivations? Body-image problems and reluctance to age. However, if you think that lifting, sucking, and filling are your only options, I'd like to offer you a far safer and more palatable solution.

As a doctor, I know that many patients who seek plastic surgery in middle age may not be receiving the minimum self-care necessary to stay healthy mentally, physically, and emotionally. Patients with high self-esteem tend to care for themselves by exercising, abstaining from drugs and smoking, eating well, and drinking plenty of filtered water—all preventive activities that encourage a long, healthy life. If caring for your health has not been your focus, I'd encourage you to ask yourself why.

Perhaps your focus has been on your children, partner, or work. Self-care requires contemplative inquiry and self-reflection. Without it, it's hard to become healthy, inside and out. One thing I know for certain is that self-care is more effective than plastic surgery.

"Never Use Age as an Excuse"

I first saw Ida Keeling on Facebook when I found a photo of her in mid-push-up, sporting a huge grin and impressive deltoid muscles. She is known for the quip "Never use age as an excuse." I was shocked to discover that she was a hundred years old. From the photo, I learned that she is a record-holder in track for the hundred-meter dash.[5] That's a lot of hundreds. There's a sparkle in her eyes despite the triple digits of her age. She demonstrated a couple of important truths about exercise: it's never too late to start (she began running at age sixty-seven), and most of us don't get the right dose (along with running, Ida bikes, jumps rope, and goes to yoga twice per week despite struggling with arthritis).

"There are people who consider themselves 'old' just sitting around at home and waiting to die—that's just stupid," Ida says. "If I could, I would tell them to stop feeling sorry for themselves and to get active, but there's nothing wrong with recharging yourself when you need to.

"Exercise is one of the world's greatest medicines," Ida explains.[6] Indeed, you may not realize that the dose of this "medicine" affects your longevity. You're probably familiar with the current recommendation of one hundred and fifty minutes per week (that's about thirty minutes five days a week). One recent study from the National Cancer Institute and Harvard University showed that for 661,000 middle-aged people followed over fourteen years, exercising at even a low dose, meaning *less than* the current recommendation, reduces mortality by 20 percent. Even better news? Bump it up to an hour to two hours per day of moderate intensity and you double the benefit.[7] That means you want to stay active as you age. In chapter 7, I will show you the exercises most supported by science for keeping you supple and I'll explain how to determine the right dose for you.

You're Not a Slave to Your Genes

The conventional wisdom is that our genes haven't evolved much since the Great Leap Forward approximately fifty thousand years ago, when a behavioral, genetic, and cognitive leap ended significant biological evolution of humans. Genetic adaptations since then have been relatively superficial; for instance, humans becoming able to digest grains and dairy, as opposed to mostly seeds, nuts, tubers, fish, fruit, vegetables, and animal protein.

Right now we are in the midst of an epoch that will revolutionize the way medicine is practiced. We're in the era of biological design. Before scientists sequenced the entire human genome, in 2003, people figured DNA was the blueprint for the cause of all disease. Quite the contrary— researchers learned that disease isn't hardwired into your DNA but is much more malleable, a result of complex interactions involving your DNA, your lifestyle, and your environment. Essentially, you have the power to reconfigure the way your DNA talks to your body, a process known as gene expression.

While DNA can't explain all your biological traits, this book addresses the key genes you can influence, the ones that affect weight, aging, appearance, stress resilience, mental acuity, and healthspan.

How to Use This Book

As you learn the backstory of your most important genes and how they interact with your lifestyle, you'll be able to turn on the good parts of your gene expression and turn off the bad parts. I'll share food, sleep, exercise, stress-busting, and brain-boosting ideas so that you can discover the top cures that will slow down your aging process. Each week of the Younger protocol is introduced in a set of practices divided by topic:

- Week 1: Feed

- Week 2: Sleep

- Week 3: Move

- Week 4: Release

- Week 5: Expose

- Week 6: Soothe

- Week 7: Think

After seven weeks, the Younger protocol works on an ongoing basis to keep cells dividing happily, sustain DNA-repair mechanisms, reduce your chance of a scary diagnosis like cancer or dementia, and decrease the likelihood that you'll need a facelift or a walker. You may want to go back through the protocol once or twice a year if your new behaviors start to slip.

Are you ready to become younger?

CHAPTER 2

The Gene/Lifestyle Conversation

Throughout your life the most profound influences on your health, vitality, and function are not the doctors you have visited or the drugs, surgery, or other therapies you have undertaken. The most profound influences are the cumulative effects of the decisions you make about your diet and lifestyle on the expression of your genes.

—**Jeffrey Bland**, *Genetic Nutritioneering*

Gasping for air, I traced Justina's footsteps in the sand, determined not to bail. We were visiting my parents at their home on the Oregon coast, and my sisters and I had gone for a run on the beach. It was a typical February day in the Northwest: about fifty degrees, blustery, and drizzling. Justina, my youngest sister, pulled away in a sprint. She loves to run fast and has the type of body that looks good in everything; I mean, she could wear a trash bag and look hot. Meanwhile, my middle sister, Anna, and I tried to keep up, both of us thinking about the delicious meal my mom was making back at the beach house. The salty air misted our skin, but we were unable to talk, and we silently resigned ourselves to watching Justina win the race.

Anna, now forty-two, and Justina, age thirty-seven, are women with whom I share roughly the same genetics, but our lifestyles and, therefore, our environmental exposures and how they affect our DNA are quite different. We're aging at different rates because genetics isn't everything.

Note the following terms that we'll be using to describe the gene/ lifestyle conversation.

- *Genetics* refers to your DNA—the study of heredity and the small variations in genes leading to inherited traits.

- *Epigenetics* concerns the interaction of genes with the environment leading to heritable changes in the way DNA is expressed in your body. The main difference between genetics and epigenetics is that the modifications are in gene expression, not in the DNA sequence itself.

- *Genomics* refers to the structure, function, evolution, and mapping of the entire genome. Genomics addresses all genes and their interrelationships in order to understand their aggregate influence on an individual.

While the science of DNA sounds like a foreign language, keep in mind that it's easy to turn to the genetics first to identify key lifestyle elements of staying young and fit.

How I Came Up with the Younger Protocol

I started testing my own and my husband's DNA in 2003, when I was considering pregnancy. We were looking for genetic problems that our daughter might inherit and, luckily, came up empty-handed. In 2005, I started testing more genes in myself. Why? I wanted to find out the ideal food plan for my body, the most expedient ways to lose weight, the best supplements to take, the most effective exercise regime, and what else I had passed onto my kids. Our genes mostly control enzymes that in turn influence micronutrients, detoxification, and metabolism. Based on the results of those tests, I added B vitamins, including methylated folate, and vitamin D to my daily supplement regimen. Then I started high-intensity interval training. *Hallelujah!* My long-standing depression disappeared almost overnight, my weight dropped, and my energy soared. I knew I was onto something important.

A few months later, not long after giving birth to my second daughter, I headed to yoga class by myself on the weekend. My husband stayed home to bond with our girls. In class, we were about to move into the peak pose, a difficult arm balance called side crane. Attempting it usually sent me toppling to one side or another. I looked down at my postpartum belly, still bigger than I wanted it to be. I could feel my breasts filling with milk; it had been a few hours since I last breast-fed my baby. Following the teacher's detailed instructions, I planted my hands on the mat in front of me, twisted my legs to the right, engaged my core, and took flight! Up went my legs to rest on a spot just above my right elbow.

My legs stayed put, and I was perfectly balanced. I was solid—and strong. I breathed slowly and smoothly. I couldn't imagine what was different and, at that moment, didn't care. I marveled at how my body had managed to retain some vestige of core strength despite the fact that I'd recently pushed a small-basketball-size baby through my pelvis. As the instructor advised us to unwind, I reluctantly came down.

Afterward, I rushed home to tell my husband. "Babe, it was amazing— side crane! My center of balance is shifted, as if my body changed for the better after having two kids. I could have held the pose forever." I paused and then said, "Wow, my body is older, better, and wiser than before!"

David responded with something like "That's great; here, feed the baby," but I knew I couldn't ignore this dramatic change. Even after all my decades of medical education, I never thought the body could get better with age. But now I had proof. It was just the spark I needed to consider the ways that my *environment*, the sum of all of my lifestyle choices, affected the way my DNA communicated within my body. Just as having a baby, trying to reclaim my body with core work, and a shift in my center of gravity had made me successful at mastering a difficult yoga pose, perhaps my patients could partner with me to determine *how to become successful at mastering the targeted lifestyle choices that would best express their DNA.*

Top Seven Genes That Affect Aging

You have about twenty-four thousand genes in your body. While many genes are important for preventing or reversing the aging of your body and mind, after years of testing my patients and creating customized protocols for them, I've found that we can narrow down your long list of genes to the top seven that matter most. I know it's probably been a while since your last biology class, so let's start with a little background. All of us possess twenty-three pairs of chromosomes, the package containing most of the DNA in your body (there's a tiny bit of additional DNA in your mitochondria that you received from your mother only). Before a cell divides, the chromosomes in the nucleus duplicate, and then the cell splits, distributing the sets of DNA evenly between two daughter cells. This process of cell division allows for growth, repair, or replacement of cells. See the figure on page 24 for the locations of these seven genes.

Overall, the human genome is about 99.5 percent the same from one individual to the next, but the 0.5 percent difference accounts for characteristics such as eye color and body shape. Each person is unique because some genes appear in different forms, called genetic variants. If a variation occurs in more than 1 percent of the population, it's referred to as a polymorphism. These variants can be categorized according to function and are due to small differences in the DNA code that change the gene in a positive or negative way. Variations result from evolution, sometimes as a result of mutations, which occur randomly in an individual and are considered an abnormal change in the gene. Certain factors—what you eat and drink, how much you sleep, how you cope with stress—can turn these genetic variants on or off.

When I tell you the names of the top seven genes, they're going to sound bizarre. Most of them seem like meaningless strands of letters and, sometimes, numbers. There's an important longevity gene that I'll describe in a moment called *forkhead/winged helix box gene, group O3* (FOXO3 for short, pronounced "fox-oh-three"). Really, that's the best name they could come up with? It makes you want to grab the scientists by the shoulders and give them a little shake. Instead, just treat the names of the genes like license plates—important, but tough to remember. Use the easier nick-names whenever possible.

Of course, there are other important genes besides the top seven. For example, I have a gene variant that makes me more likely to lose weight when I eat more fish than meat. It's called PPARγ, which stands for "peroxisome proliferator-activated receptor gamma." It controls how the body responds to certain types of fat. One study showed that women who have this gene lose weight when they consume more than 50 percent of their fat from the omega-3s and omega-6s found in fish, shellfish, and nuts.[1] So by eating more fish and nuts, I lost weight, and you can too (if you have the same gene). You'll learn more about this gene in chapter 5.

In the top seven, I've included the genes that come up repeatedly and have the greatest impact on your healthspan when targeted by lifestyle changes. In other words, they have the strongest gene-environment inter-action. As an example, I am fortunate to have inherited normal versions of the genes BRCA1 and BRCA2; the abnormal or variant versions may cause a higher risk of breast cancer. (Other women in my family weren't as lucky as I was.) An example of one of my *variant* genes is Fatso, described below, which makes me more likely to be hungry, unsated by food, and fat compared with someone with a normal Fatso gene.

I will help you determine whether these seven genes are working for or against you. With each week of the Younger protocol, you'll learn how to turn on and off each gene as appropriate to balance the way these seven genes are expressed, which may help thwart the aging process.

1. Fatso Gene

Official name: Fat mass and obesity associated (FTO) gene

Location: Chromosome 16

Job: This gene is strongly associated with your body mass index and, consequently, your risk for obesity and diabetes. When you have the vari-ant, it gives you sloppy control of leptin, a hormone in charge of satiety. In other words, you're hungry all the time.

Your task: You can turn off the Fatso gene with exercise and a low-carbohydrate food plan that's high in fiber.

2. Methylation Gene

Official name: Methylenetetrahydrofolate reductase (MTHFR) gene

Location: Chromosome 1

Job: The Methylation gene provides instructions for making an enzyme that plays an important role in the processing of vitamin B_9 and amino acids, the building blocks of proteins. It also helps you detoxify alcohol.

Your task: Work around a defect in the Methylation gene by eating adequate folate—not too little, which may lead to depression, high blood pressure, heart disease, stroke, addiction, and cancer.

3. Alzheimer's and Bad Heart Gene

Official name: Apolipoprotein E gene

Location: Chromosome 19

Job: The APOE gene instructs cells to make a lipoprotein that combines with fat and transports cholesterol particles in the blood and brain. People with the bad variant of this gene, APOE4 (or sometimes APO-e4), don't recycle cholesterol, leading to higher levels of low-density lipoprotein (LDL, or bad cholesterol) in the blood. Women with APOE4 have a threefold greater risk of developing Alzheimer's disease.

Your task: When the good variant (APOE2 or APOE3) is turned on, you may lower your risk of heart attack, stroke, and Alzheimer's disease. When you have one or two copies of the bad variant (APOE4), you will want to turn it off with the strategies in the Younger protocol, such as sticking to an anti-inflammatory diet, exercising, keeping your blood sugar stable, and getting restorative sleep.

4. Breast Cancer Genes

Official names: BRCA1 (from "BReast CAncer gene one") and BRCA2 (from BReast CAncer gene two") genes

Locations: Chromosome 17 (BRCA1) and chromosome 13 (BRCA2)

Job: The BRCA genes belong to a class of tumor-suppressor genes that repair cell damage and breaks in DNA and that keep breast cells growing normally. When you inherit the variant, you may not be able to prevent breast tumors from forming. Overall, one in four women with breast cancer are known to have a gene variant. There are thousands of variants of these breast cancer genes, and there are probably another hundred other breast cancer genes (such as TP53, PTEN, CHEK2, ATM, and PALB2).[2] Even for women with BRCA1 and BRCA2, the range in risk is broad: 20 percent in some people, and 90 percent in others, which means out of one hundred women with the BRCA1 or BRCA2 mutation, somewhere between twenty to ninety will develop breast cancer over their lifetimes. Without intervention, a woman with a BRCA gene *mutation* is seven times more likely than other women to get breast cancer (and thirty times more likely to get ovarian cancer) by age seventy.

Your task: Turn off the breast cancer genes by eating more vegetables and less inflammatory meat, drinking less alcohol (no more than one serving twice per week), and keeping your inner clock ticking at a normal pace.

5. Vitamin D Gene

Official name: Vitamin D receptor (VDR) gene

Location: Chromosome 12

Job: VDR codes for the nuclear hormone receptor for vitamin D_3, which enables your cells to absorb vitamin D. When you inherit the variant, you are more likely to suffer from bone loss.

Your task: If you have a bad variant, like I do, you need to open the vitamin D receptor by keeping your vitamin D blood levels even higher than recommended by conventional doctors, with a target range of 60 to 90 ng/mL. My vitamin D receptor functions at half the level of a normal VDR, so I keep the amount of vitamin D in my blood at about double the recommended level in order to work around my bad variant. Put another way, your task may be to raise your intake of vitamin D beyond the standard recommendation of 1,000 to 2,000 IU per day.

6. Clock Gene

Official name: Circadian locomotor output cycles kaput gene

Location: Chromosome 12

Job: This gene modulates circadian rhythm, or the twenty-four-hour biological sleep-wake cycle. If you have the bad variant, you will have higher blood ghrelin levels (the hormone that makes you hungry) and resistance to weight loss. Other hormones released on a circadian clock will be affected.

Your task: Protect your circadian rhythm, keeping your body on a normal sleep-wake cycle, one of the most important regulators of your hormone production. You must get the right amount of sleep to lose weight if you have the bad variant of this gene.

7. Longevity Genes

Official names:

- Mechanistic target of rapamycin or mammalian target of rapamycin (mTOR) gene or, if you prefer, FK506-binding protein 12-rapamycin-associated protein 1 (FRAP1) gene

- Forkhead/winged helix box gene, group O3 (FOXO3)

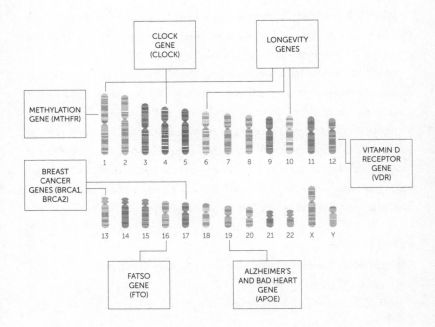

- Sirtuin, called SIRT1, which protects you from diseases of aging by revving up the mitochondria, the power plants inside cells that tend to conk out as you get older

Locations: Chromosome 1 (mTOR), chromosome 6 (FOXO3), and chromosome 10 (SIRT1)

Job: Your longevity genes regulate cell growth, proliferation, motility, survival, and protein synthesis. Some variants are associated with a shorter life; others with a longer life.

Your task: Switch the longevity genes to a longer healthspan, which is sometimes performed differently for each gene. For instance, sitting in a sauna for twenty minutes turns on the FOXO3 longevity gene. Intermittent fasting turns on SIRT1 and turns off the mTOR gene; when hyperactive, mTOR is associated with Alzheimer's disease, cancer, and early mortality.

Timekeeping Telomeres

When your body is aging fast, there are several telltale signs:

- The scale creeps up every year.

- You are foggy after sharing a bottle of wine with someone.

- Your frown lines no longer disappear when you smile.

- You spend more time than you used to looking for your keys.

- The discs between your vertebrae shrink, and you may develop low back pain and stiffness.

- Even worse, you may be facing a diagnosis of disease.

The extent of these changes are actually measurable in your blood by way of a test for *telomere length*. You have timekeepers in your cells called telomeres. A telomere is a collection of DNA at the end of your chromosome that acts like a knot at the end of a thread. A telomere signals to the enzymes duplicating your DNA that they're almost at the end of the strand and it's time to stop, just as a knot tells you to stop pulling a piece of thread with a

You Don't Have to Test Your Genes

With all this talk about your DNA, you may wonder if you need to test your-self to benefit from this book. In a word: *no*. As I mentioned, 99.5 percent of human DNA is identical from one individual to the next, so maximizing the way your DNA functions is very much the same for all of us when it comes to aging. Furthermore, you have a ton of genes! In one study that looked at 2.8 million genes in 320,485 individuals, one hundred genetic variants were found to contribute to BMI, and they don't change as you get older.[3] The reason I've chosen the Fatso (FTO) gene is that it has the largest impact of all the hundred genetic variants—in other words, the Fatso gene holds the greatest potential for change in your body.

Another important factor is that only 10 percent of disease is caused by your genes; 90 percent is caused by your environmental factors, which may turn genes on and off. Therefore, in the seven weeks of the Younger protocol, our focus is on how to upgrade that 90 percent to affect the genetic 10 percent. The foundation of this protocol consists of the most proven steps to improve your environment and change how your DNA is expressed—literally, how to turn on and off your genes.

Perhaps the most important reason that you don't need to test your genes is that genetic testing is not 100 percent accurate. Even the most commonly performed tests may be off because of the orientation of the gene, which is sometimes read in the forward direction and sometimes backward on the chromosome.[4] That means the results of genetic tests should be taken in the context of your particular risks and reviewed with a knowledgeable health professional who understands the interplay be-tween genetics and the environment and limitations of specific tests.

If you decide to test, it is increasingly more affordable. At the time of the writing of this book, it costs about two hundred dollars to map important genes (see Resources). In the future, I predict most of us will be running around with our own genomes printed on smart cards kept in our wallets. This will allow for a more personalized approach to preventing disease and unnecessary aging. Until that day arrives, the Younger protocol can pro-duce antiaging results even without genetic testing.

needle. For nearly all normal cells, each time a cell divides, its telomeres get shorter. At a certain point, the cell dies because the telomeres are gone and the ends of the chromosomes are no longer protected. It's normal to lose telomere length as you age, but only at a certain, healthy rate. Some people lose length on their telomeres faster than average.

Short telomeres put you at a greater risk not just for wrinkles but also for heart disease, cancer, and early death. Overall, people with short telomeres have an up to 300 percent higher risk of cancer of the pancreas, bone, bladder, prostate, lung, kidney, and neck. Thankfully, you can improve the upkeep of your telomeres. When you do, you will look and feel younger than you have in years. In the following chapters, you will learn the secrets of my patients who have the telomeres of women ten to twenty years younger, despite stressful lives.

Think back to those celebrities we admire—they are fully aware of the $90/10$ rule. Genes are only 10 percent of the story of why these women keep their enviable body structure as they age; the other 90 percent is determined by their lifestyle and how it affects their biochemistry and subsequently expression of their genes, an interaction known as epigenetics. These women apply their wealth to help out their genes by paying for the top personal trainers, private chefs, and nutritionists to keep their good genes turned on and their bad genes, such as the ones for weight gain and breast cancer, turned off. Most work hard to keep their A-list bodies; they eat fish almost daily, consume eggs from their own organic chicken coops, and love to chomp on dark leafy greens. When they eat a chocolate chip cookie or drink a glass of wine, they stop at one. Many practice yoga three times per week, kickbox, and diligently complete their cardio workouts. Instead of applauding their platinum genetics, we should applaud their platinum epigenetics.

Think about epigenetics as the blueprint for a house. If you've built or remodeled a home, you know there's an initial blueprint, but it gets revised during the design-and-build process. The final home is rarely exactly the same as the initial blueprint. So it is with your body. You were conceived with your initial blueprint, or DNA, but it probably got modified before you were born by what your mother ate or even by what your grandmother ate. After that, it was affected by whether you were born vaginally or via caesarean section, if you were breast-fed, when and how often you took antibiotics, and other environmental factors. The modifications to your blueprint are *epigenetic*—that is, the non-DNA biochemical changes that alter the way your DNA is expressed.

Disparate Sisterhood

Here's a look at my own family that will help us better understand the role of epigenetics.

I'm the oldest child, born in 1967. My mom was very thin; she weighed about 120 pounds at a height of five foot seven inches. Breakfast for her was coffee on the run, a quick sandwich for lunch, and maybe a small serving of meat and potatoes for dinner. She probably under-ate while pregnant with me and gained twenty pounds, which ironically programmed me to be fat and have blood-sugar problems as an adult (as I'll describe later).

Starting with my time in utero and into my early childhood, my mother's behavior put a tag (like a sticky note) on my genes in locations that determine weight and blood-sugar control. It's as if I'm famine-proof: I can't lose weight even if I starve myself with restricted calories. The sticky note on my genes announces the following to the rest of my body: "Team, she's depriving us of food, so let's make sure she doesn't starve to death. Brain, get her to think about food all the time and make her binge whenever possible. Thyroid, slow down the rate you burn calories and store it all over the body in case we need it. Belly fat, stay where you are—it could be a long haul, so don't burn off any fat, no matter what." The ability to survive a famine with little food intake is great for human evolution and keeping me alive but not so great for rocking my little black dress.

Breast-feeding wasn't trendy when I was born, and because my mom worked full-time, she breast-fed me for only two months. (What your mom ate while pregnant with you and the length of time you were breast-fed affects the establishment of your gut flora and their DNA, which are key factors in your overall health, among other things.) My grandmother picked me up from school and took me home to snack on cookies every day, and I'd finish my homework while watching cartoons and drinking milk.

If you've read my first two books, you know I was never really an athlete and I tended to be overweight until my thirties. Since I discovered a protocol to reset the hormones of metabolism, I've stayed at a relatively healthy weight of about one hundred and sexy pounds at five feet six inches. Keeping my weight in the ideal range has become more difficult, to put it mildly, as I get older, and it's sometimes a ridiculously enormous project. As a recovering stress case with a tight neck and shoulders, I work diligently to keep my blood sugar, thoughts, and weight under control, although sometimes the number on the bathroom scale amplifies my stress, and it might do the same for you!

My middle sister, Anna, loses weight easily when she sets her mind to it, as she did after the birth of her son. Mercifully, she is as tall as my mother and has long legs. Mom gained about forty pounds when she was pregnant with Anna. Mom had the time to breast-feed Anna longer—about thirteen months—and took six months' leave from work and then went back part-time. Anna was an active child, involved in volleyball and track during school, although now as a teacher and working mom, she finds it tough to fit in exercise.

Like me, she is a recovering stress case but she doesn't have the famine genes turned on the way I do. We both are in caregiving professions that squeeze us emotionally dry at times. Her response to stress used to include a glass or two of wine and maybe chips and guacamole, but now she makes phone calls to friends and no longer drinks. I channeled my stress into yoga, and I've become a yoga instructor so I can better balance my life. Recently, Anna dropped forty-five pounds by following the tenets of my second book, *The Hormone Reset Diet*, and has easily kept it off. Was it my mother's in utero programming and postpartum breast-feeding that helps her stay svelte more easily than me? Perhaps.

My youngest sister, Justina, is the prettiest, most well-adjusted one. Justina was born in 1979, around the time that my mother evolved into a serious foodie. Mom had become an advocate of Alice Waters and the organic locavore movement in the San Francisco Bay Area. We ate free-range-egg omelets with greens from our organic garden for breakfast.

Mom gained twenty-five pounds with Justina and breast-fed her for six months before Justina went on strike and refused to continue despite my mother's fierce commitment. Mom took several years off from work to be at home with Justina. She was a healthy and sweet baby, and she must have gotten my father's jock gene because by age five she was already developing into a soccer prodigy, and she later played at the college level. The muscle memory of her early and consistent athletic training has stayed with her throughout adulthood, and she has not gained or lost more than a few pounds since high school. She has wisely chosen a simpler life in rural Oregon with her husband and her dog, a gigantic mastiff, which leaves little room for stress to overtake her the way it does Anna and me. While Anna and I commiserate over our latest stressors, Justina doesn't really understand what the fuss is all about. Justina deflects stress with a daily run on the beach and a powerful capacity to shrug off the small stuff with humor, grace, and time with the dog.

You may wonder if the disparity between my siblings and me is just a difference in body type, but many times both Anna and I have been at Justina's weight and had the same body shape. The explanation comes down to our different *environmental inputs*, ranging from my mom's healthy cooking while pregnant with Justina, regular athletic training (which perhaps also contributed to a degree of stress resilience that sometimes evades me), and her ability to deflect stress.

You and your siblings may have far more discernible and dramatic differences, but the point is the same. You can't change what your mom did while you were in utero and as a child; you can change your current and future environmental inputs. You can build muscle and make stress an ally. You can retrain your mind and your fork, and do it without deprivation so that your famine genes stay off. You can learn to love matcha or organic wine so that your longevity genes stay in the on position.

The Paradox of Survival

Your body has limited resources, and that means you make judgment calls every minute without realizing it. Body maintenance—removing mutated DNA, improving your enzyme production when it becomes sluggish, getting rid of damaged proteins, neutralizing highly reactive molecules called free radicals—takes work but it's the way to slow down aging. How fast or slow you age is the product of how well your body sweeps up the damage of daily living. Ultimately your body must decide where to send its limited resources: reproduction, growth, physical work and exercise, and/or repair and maintenance.

An important example is the famine genes that I mentioned, which may have evolved to help people survive long periods without food. People who have their famine genes turned on, such as the Irish who outlived the potato famine or Ashkenazi Jews who survived pogroms in Eastern Europe, are gifted at banking fat. They stay alive during times of hardship, when food is scarce. Fast-forward to modern life and our surplus of food; the genetic tendency to bank fat starts to work against them. The very genes for insulin resistance that allowed them to survive a famine now makes them chubby, no matter what they try. Just because the famine ends doesn't mean the genes switch off. The key is to understand the workings of the famine genes (if you have them, as not everyone does) and to override them (i.e., turn them off) so you can remain lean even when food is plentiful.

Another example is reproduction genes. The genes that help you grow and reproduce are at odds with the genes that help maintain and repair your cells—almost like a double cross later in life. Consider a man with high testosterone at age thirty. He has a better chance of impregnating a woman than a man who has lower testosterone does, but the man with lower testosterone will live longer. As we navigate the Younger protocol, we need to be aware of these paradoxes. The goal is to avoid an early death and lengthen healthspan. So the key is to turn on the right genes at the right time and in the right sequence. Here's what we'll cover.

Week 1: Feed. You will start by applying the counterintuitive but easy-to-follow directions that control the interaction between genes and what you put in your mouth: food, drinks, and supplements. We'll address the longevity and vitamin D genes, turning off the Alzheimer's and Bad Heart gene, and regulating the Fatso gene and your metabolism. The focus will be on the actions that enable your body to produce enzymes, hormones, and other substances essential for turning off the ticking time bombs in your cells.

Week 2: Sleep. In this chapter, you'll learn how to get your Clock gene working for you, even when you are super busy, ruminate, or can't sleep through the night. If you're like me and have the variation of Clock, you need eight hours of sleep per night to lose weight, because the gene variant can raise your daily levels of ghrelin, a hormone that makes you hungry.

Week 3: Move. Ever heard of the sitting disease? You'll learn how to combat it by turning on thousands of good genes via exercise. You'll discover which forms of exercise are best for outsmarting the aging process, preventing cancer, enhancing mental health, renewing skin, and turning off the Alzheimer's and Bad Heart gene. You'll become skillful at managing overload when it comes to your form of exercise and the role of intensity. I'll help you find the right dose.

Week 4: Release. Holding tension and habitual patterns of tightness in the body is the early stage of stiff joints and muscles, later leading to reduced mobility and slower gait. In the ancient yoga tradition, activating *bandhas*, or energetic locks, in your body can delay aging. Learn the self-adjustments, *bandhas*, and other techniques for releasing your restricted pathways and increasing your mobility so that muscles can keep working for you. You'll turn off genes that make you injury-prone, such as the Achilles gene, and activate longevity genes.

Week 5: Expose. You will learn about the genes that tend to make the biochemistry in your body go sideways, such as the Methylation, Breast Cancer, Vitamin D, Skin and Wrinkle, and Mold genes. You'll learn the most proven positive environmental exposures that can turn on your longevity genes, improve your skin, and optimize your immune function.

Week 6: Soothe. This week, you'll learn how to turn off the genes that make your response to stress more easily activated and/or make it harder for you to return to baseline. You will discover proven ways to mend your timekeeping telomeres. You'll also learn how to turn on the Bliss gene, and who doesn't want that?

Week 7: Think. You will focus on shifting the balance toward more remembering and less forgetting using proven strategies in functional medicine. You'll turn off the Alzheimer's gene and enhance your self-talk so it's more loving, and you'll retrain your mind to minimize cognitive distortion. You will discover which supplements improve your cognition. Did you know that having low levels of vitamin D more than doubles your risk of dementia (sometimes called vitamin-D-mentia)?[5] With your vitamin D levels in the optimal range, you can preserve your good brain.

At the end of the seven-week protocol, you'll integrate key aspects in order to *sustain* your healthspan. This is the most important week, where you learn how to maintain the progress you've made. Make the Younger protocol your new gold standard for how you take care of yourself and lengthen the best of middle or older age.

Epigenetics may be the key to extending your healthspan. This is the promise of personalized-lifestyle medicine. Death is inevitable, but your healthspan is up to you.

CHAPTER 3

Epigenetics: Turning Genes On and Off

We now know that the silencing and activation of genes using various chemical tags and markers is a pervasive and potent mechanism of gene regulation. The transient turning on and off of genes has been known for decades. But this system of silencing and reactivation is not transient; it leaves a permanent chemical imprint on genes. The tags can be added, erased, amplified, diminished, and toggled on and off in response to cues from a cell or from its environment.

—Siddhartha Mukherjee, *The Gene: An Intimate History*

M y best friend from high school, Natalie, is French. She and I traveled to France at the end of senior year and visited her eleven aunts and uncles in various locations from Paris down to Toulouse and Nice. I mention this not because I'm about to regale you with stories of the delicious food or amazing landscapes; what I recall most vividly is how thin the women were despite the fact that they drank bottles of wine and ate chocolate croissants, loads of cheese with baguettes, and potatoes fried in duck fat. I come from peasant stock in Ireland and Germany, so I don't have the thin French body type. But I discovered later that it's not really a body type or what the French eat that matters most; it's a particular gene type mixed with a particular lifestyle.

Even now that we're fifty, my friend Natalie is not fat and never has been. Like me, she's a mother and has a full-time job, but she's a normal eater, and an extra glass of wine doesn't show up on her hips like it does on mine. We get about the same amount of exercise, calories, and chardonnay, but Natalie looks better.

It's true that far fewer French women get fat, but the reason has less to do with the shared bottle of wine over dinner and more to do with its interaction with the Methylation gene, methylenetetrahydrofolate reductase (MTHFR). This gene determines how chemicals are tagged in the body, or methylated, and also how you detox alcohol in your system (alcohol blocks methylation). Siddhartha Mukherjee in *The Gene* refers to methyl tags as decorating the DNA strand, like charms on a necklace, thereby quieting the gene.[1] Specifically for our discussion, the MTHFR gene codes for the MTHFR enzyme, which provides instructions for how to make vitamin B_9 usable by the body. I have about a 35 to 40 percent reduction in the activity of my MTHFR enzyme because I have a variation of the MTHFR gene—I inherited one normal copy of the gene from one parent, and one variant copy from the other. There are three potentially serious problems linked to my mutated gene: I can't make sufficient vitamin B_9, I don't detoxify alcohol well, and I don't convert the amino acid homocysteine into methionine, an important constituent in muscle growth, among other things.

I learned early on that I must drink less alcohol than Natalie and eat more folate, which is found in dark leafy greens (turnips, collard, mustard) and other vegetables (asparagus, spinach, romaine lettuce, broccoli, cauliflower, beets). When you know that you're predisposed to be a poor methylator and detoxifier of alcohol, you can change your inputs to create balance in the body. You can eat more greens and drink less wine as a *genetic work-around*. As you'll learn in this chapter, environmental inputs like alcohol and folate from dark leafy greens may exert a greater impact on your body than your genes do.

You definitely have gene variants too—that's how your ancestors evolved and managed to pass on their genes to you. I've run genetic tests on thousands of my patients, and everyone has at least three scary genetic mutations. Then again, mutation doesn't mean dysfunction. Genes code for proteins (usually enzymes) most of the time, and a mutation simply means that you make more or less of that protein. There's no need to throw up your hands in defeat; you merely want to know how to work around your particular genetic mutations—that is, how to hack your environment so that you can live long and be strong.

Hacking Your Environment

I love the axiom that genetics loads the gun, and the environment pulls the trigger. Genetic factors and environmental factors don't work on their own—they interact. Even if you have lousy genes, you can change your health destiny by managing your exposures. It means that you have the power to get your body to work for you instead of against you. You can actually adjust epigenetics up or down by tweaking what's known as your *exposome*, a collection of environmental factors that directly or indirectly affect your health. You control your exposome by your daily habits of body and mind, both conscious and unconscious, including how often you move and what form that movement takes, what environmental exposures you have in your home and office, what you eat and drink, and how you manage or mismanage your hormones. When one gene is turned on by a certain exposure, another gene may be turned off. We want to optimize the collective effect in the Younger protocol, tuning the immune system (and other bodily systems) by just the right amount so that your body has an excellent defense team.

Think of these aspects of your health and the aging process as concentric rings:

- At the center is your DNA, your blueprint, which is determined by your parents.

- Next is your exposome, non-DNA activity that influences whether genes are switched on or off, akin to change orders.

- Which genes are switched on and off then determine health conditions you face as you age, such as weight gain, wrinkles, and low energy.

- If left unchecked, these conditions may become diseases, such as diabetes, Alzheimer's, and obesity, and might lead to premature death.

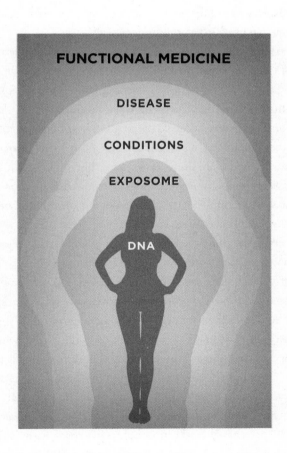

DNA Is Slow to Change, but Gene Regulation Can Change Fast

New scientific breakthroughs provide important clues about how to live younger and longer, suggesting that you can embrace certain behaviors to boost your positive environmental cues. While DNA change is slow, regulation of gene expression can change more rapidly—sometimes temporarily, and sometimes permanently. Modifications of gene regulation that activate or deactivate the way a gene is expressed may be heritable, known as epigenetic changes, so that you may pass on your good or bad exposures to your offspring.[2] In short, you want to direct your gene regulation not only for yourself but also for your progeny.

Imagine identical twin brothers with the same genetic blueprints. One becomes a type A investment banker and ultramarathoner, drinks coffee every morning and bourbon every evening, and hardly sleeps despite a nightly Ambien. The other moves to Tibet, becomes a monk, and meditates five or more hours per day. The first twin will have a faster metabolism, higher stress load, a shrinking brain from the alcohol, and poor recovery due to the lack of sleep. He will most likely die sooner. While you don't need to become a Tibetan monk to slow down aging, scientific truths from these two extremes can be translated into a meaningful protocol for slower aging. Every day is a fresh chance to be younger. That's the thrilling promise of epigenetics, and it goes far beyond the commonsense strategies of eating more vegetables and walking in nature.

Most of your choices and habits represent an incredible opportunity for scientists—and you—to prevent and reverse disease. For example, you may not have a family history of breast cancer; however, if you have an imbalance of the good and bad bacteria in your gut, you may make more of the dangerous and provocative estrogens that elevate your risk, and you may make fewer of the protective estrogens that lower your risk. As a result, you keep recycling estrogens and overstimulating estrogen receptors, potentially increasing your chances of developing breast cancer.[3] Keep in mind that 85 percent of breast cancer occurs in women with no family history, so please don't falsely reassure yourself if your mother, grandmother, and aunt are disease-free. Your gut may be working against you, and you may not even know it.

By making lifestyle changes, such as reducing your alcohol consumption, exercising more, and losing weight, you can potentially reprogram a gene to tell your body to make more of the "good" estrogens instead of the "less good" estrogens.[4] Overall, there's no change in your DNA sequence, but nongenetic triggers can cause your genes to behave differently.

Preventing Breast Cancer Two Ways

In 2013, Angelina Jolie published an opinion piece in the *New York Times* describing what she'd done when she found out she carried a faulty gene, called BRCA1, that gave her up to an 87 percent chance of developing breast cancer and a 50 percent risk of developing ovarian cancer.[5] Her mother, grandmother, and aunt also presumably had this gene and sadly had lost their battles to cancer. So at age thirty-seven, Angelina had chosen to undergo a prophylactic mastectomy to surgically remove both breasts. Two years later, she also chose to undergo prophylactic removal of her ovaries.[6] This is one rather costly and extreme way to prevent breast and ovarian cancer, and since only 15 percent of women have a family history like Angelina, the rest of us have to consider other options for prevention that are less expensive and more palatable—just like Marie.

Marie was sixty-six when she found a drop of blood on her white bra. While no one in her family had breast cancer, she called her gynecologist, who sent her for an ultrasound. It took forever, but the radiologist found a tiny nodule. They decided to remove the nodule, and the biopsy came back as *atypical hyperplasia* of the breast. In other words, Marie had an accumulation of abnormal cells in the breast. It wasn't malignant, but it meant her risk of breast cancer jumped up fourfold, which is enough to scare the pants off any woman.

Her breast surgeon told her that taking the anti-estrogen drug tamoxifen would help her prevent breast cancer, and she reviewed the risks and benefits of the drug. They included a higher risk of endometrial cancer, which hardly seems like a good tradeoff. That's when I got involved; Marie came to see me for a second opinion. My suggestion? "Start by eating more vegetables—about two pounds, or ten cups, per day—begin taking a daily greens powder, cut back to two glasses of wine per week, lose twenty-five pounds, and eat less conventional red meat. Cut out inflammatory foods like dairy, sugar, and gluten. Meanwhile, we need to look at how your body produces and gets rid of estrogen to see if we can nudge the process in a better direction."

More Caveats on Genomic Testing

DNA tests are almost as readily available as blood tests. But just because you *can* get them doesn't mean you should—at least, not yet.

At the time of the writing of this book, direct-to-consumer DNA-testing companies like 23andMe offer limited carrier status information for thirty-six conditions because of a regulatory injunction by the Food and Drug Administration, which stems from questions about the validity of the tests—such as their false-positive and false-negative results—and ways that clients might misinterpret or otherwise misuse the data. We don't want people testing their BRCA genes and making a rash decision about prophylactic surgery (something known as the Angelina effect).

Another problem is that most physicians have no idea how to decode genetic-test results and provide meaningful counsel. Therefore, it's important to put into context what DNA testing is and what it isn't. A DNA test won't tell you how you'll die or even what will make you sick. It will give you clues about how to design your lifestyle and optimize function for a lower probability of disease and a longer healthspan. At some point in the future, the tide will turn and genetic testing will become more essential.

Some view the FDA's regulatory action as a restriction of personal freedom. If this is your point of view, and you want to test, then I'd encourage you to do it with a fully validated DNA panel (see Resources) that has the science to support the results. In addition, ensure that you can review the panel with a well-qualified clinician who is aware of the limitations of genome-based tests.

Six months later and twenty-seven pounds lighter, Marie went to see her breast surgeon, who said, "None of my patients did what you've done. How did you do it?" The surgeon spoke of other patients who were obese and experienced breast cancer recurrence and told her how heartbreaking it was to have to break the bad news. She noted how frustrating it was to try to help women make the lifestyle changes needed to lose weight, which would reduce their risk of breast cancer and early death.

Isn't this the trouble for all of us? We aren't making the connections between the glass of wine every night, the inflammatory fats we eat at restaurants, the short shrift we're giving to sleep, and how all of this creates a neighborhood that may foster breast cancer.

After checking the estrogen levels in her urine, I suggested that Marie add a new supplement called di-indole methane (DIM), which is an extract of cruciferous vegetables. Taking one pill is like consuming twenty-five pounds of broccoli, which helped her make more of the good estrogens that protect her from breast cancer and fewer bad estrogens that raise her risk of it.

Now Marie is more conscious about making healthier choices in all areas of her life. Nothing crazy—she eats more vegetables, walks briskly three times a week, wears a tracker to measure her steps, goes to yoga weekly, and gets a massage once a month. It was no surprise that at her subsequent exams her breasts looked healthier than ever—less dense and no signs of atypical hyperplasia on her breast MRI, performed every six months. She's kept the extra weight off for seven years. Both Angelina and Marie faced a higher risk of breast cancer, yet both chose preventive approaches that were very personal and extremely different. These two cases reveal how the groundbreaking science of genetics and epigenetics offer us more choices to prevent illness and improve healthspan.

The Basic Genetic Terms to Know

The way that your genetic variations are expressed is exactly what we want to affect with the Younger protocol: we want to turn off the bad gene variants and turn on the good. I've included a short list below of the terms you need to know to make sense of this book (and I've included a list in the glossary, located in the appendix). Focus on the terms that I'll repeat: *gene, DNA, allele, variant*. No need to get overwhelmed; just refer to the glossary if you're unsure about a term when it comes up later in the book.

DNA by the Numbers[7]

- You have forty-six chromosomes—twenty-three from your parents—each containing tightly coiled DNA.

- You have about twenty-four thousand genes, so keep that in mind when I refer to the top seven. Many genes overlap and even oppose each other in function, and it's the aggregate result that matters.

- Most of your DNA (99.9 percent) is in your cell nuclei, but you also have some DNA (0.1 percent) in your mitochondria, a separate organelle within each cell. (Your body is made of fifty trillion cells.) This is important to know because mitochondria tend to poop out, and that's what may make you tired over the years.

- Your DNA, which is an abbreviation for deoxyribonucleic acid, is made up of repeating patterns of four chemical bases: adenine (A), cytosine (C), guanine (G), and thymine (T). These bases are the alphabet of your genetic code, or genotype. Your DNA is like a ladder, with the base pairs forming the ladder's rungs. (The sides of the ladder are composed of sugar and phosphate.)

- The ladder spirals into what's known as a double helix so that it can be efficiently copied when cells divide; if you unwound your DNA, the entire ladder would be six feet long.

- About one in three hundred nucleotides has a variation (known as a polymorphism). Overall, you have about ten million polymorphisms.

- You inherit one copy of a gene from your mother and one copy from your father. The single copies are called alleles. If you inherit the normal copy of the gene from each parent, you are considered "wild type," or normal. If you inherit one copy of the normal gene and one copy of the variant gene, you are heterozygous for the gene. If you inherit two copies of the same polymorphism, you are homozygous for that gene. Problems arise mostly when a person is heterozygous or homozygous.

DNA Whispering

Five cultures in the world are famous for the longest-living residents. They have certain habits in common that switch on the right genes and switch off the wrong genes when it comes to aging, which results in the people in these cultures living twelve years longer than the average person in the rest of the world. Think of it as DNA whispering. They all live on the coast or in the mountains. They eat fish. They consume fresh food in season. They dine on a particular superfood that is rich in antioxidants, such as seaweed in Okinawa, Japan, where women live the longest, or olive oil and red wine in Sardinia, Italy, where men live the longest. These cultures arrived at a particular combination of genes and lifestyle that protects them from the ravages of aging. And so can you.

Remember, you don't have to know your DNA in order to improve your epigenetics. Many of the techniques for slowing down aging will work for you if you follow the seven-week protocol and create an exposome surrounding your DNA that will activate your best healthspan.

Changing the Way Genes Talk

The turning on and off of genes is called gene regulation, and our understanding of the various processes and how they fit together is still evolving. Your lifestyle choices influence gene regulation—you control it indirectly by eating certain foods, among other actions, as you'll learn in chapters 5 through 11. Your goal is to make sure these processes are working to the best of their ability. In this book, we'll focus on the role of transcription factors, methylation, and newer but less proven methods such as histone modification and chromatin remodeling. Other scientists may describe the hierarchy of epigenetic mechanisms in a different way as this young field continues to unfold.

While those may sound like the stuff of Nobel Prize–winning research, I want to describe a few gene regulation processes in a way that's easy to understand. Here's a quick summary, so we're all on the same page, but you don't need to understand these to benefit from the seven-week protocol.

Transcription factors. Transcription factors are proteins that can turn specific genes on or off by binding to a DNA sequence and controlling the rate of transcription of DNA into messenger RNA. Examples are heat shock factors released by taking a sauna which upregulate the genes that help you survive at high heat,[8] carbohydrate-sensing transcription factors that are involved in blood sugar control,[9] or the estrogen receptor transcription factor.[10]

Methylation. Methylation is a simple biochemical reaction that tells your cells what to do, usually by turning off a gene. This process occurs a billion times per second in every cell of your body, and that means it's important and tightly regulated. Methylation is when your body tags a strand of DNA or a vitamin with a work order to turn off a gene or a part of a gene. It involves attaching a methyl group (made up of one carbon and three hydrogen atoms) to a molecule, usually a protein or enzyme. You need methylation to inactivate estrogen so it doesn't keep circulating and causing harm. Methylation helps produce glutathione, a major force in your detoxification cycle and arguably your most powerful antioxidant. You require glutathione to rid the body of mercury when you eat sushi and to bolster your ability to clear mold from your system when you're in a water-damaged building. If your body doesn't methylate well, you won't be able to detoxify well—potentially making you more susceptible to heavy-metal toxicity and more likely to be harmed by other poisons, like pesticides, environmental toxins and pollutants, and lipopolysaccharides (a toxin you encounter when you have bad bacteria in your gut). In general, a greater susceptibility to toxic overload happens because the body is not able to detoxify properly. See sidebar for signs that your methylation is suffering, which I find in approximately 70 percent of my patients. As an example, I'm a poor methylator, so toxic estrogens tend to build up in my body. When I started eating greens daily and taking supplements that aid my methylation, I inactivated most of my bad estrogens and now my body knows to remove them in my urine and poop. (Crass, but true!)

Histone modification. There are proteins in chromosomes called histones that are not part of the genetic code but that act as spools around which the DNA winds. Histones have some control over whether genes are switched on. In general, there are several ways to modify histones: methylation, adding an acetyl group, and phosphorylation, among others. Adding an acetyl group will turn on a gene; the acetyl group is a like an *activation switch* that turns on when it's attached to a histone (biochemically, it's C_2H_3O).[11] The acetyl group is added by a histone acetyltransferase, which is like a little taxi that drops off the acetyl group to be added to the histone. Researchers found that naturally occurring changes to these proteins, which affect how they control genes, can be passed down from one generation to the next and so influence which traits are inherited. The finding demonstrates that DNA is not solely responsible for how characteristics are inherited. This discovery paves the way for research into how and when this method of inheritance occurs and whether it is linked to particular traits or health conditions.

In the introduction, I mentioned that one of the five factors that contribute to aging is that as you age, you accumulate more fat and you lose muscle. This unfortunate process occurs commonly in the liver, leading to a condition known as fatty liver or, more technically, nonalcoholic fatty liver disease, which is associated with histone acetylation. The trigger is eating sugar, which prompts acetyl groups to turn on the genes that store fat.[12]

Chromatin remodeling. Chromatin is a combination of DNA, proteins, and RNA that's found in cells. It serves to package DNA into a tiny volume that fits in the cell when all zipped up, with DNA wrapped around histones, and it allows another form of control over gene expression and DNA replication. Don't worry about the details; the main point is that it's a separate way that your body can control gene expression by revising nucleosomes using special enzymes. Certain neurological disorders, including autism spectrum disorder, are related to problems with chromatin

Signs of Impaired Methylation

- Fatigue—either low or inconsistent energy
- Poor exercise tolerance
- Obesity and weight gain
- Pain, such as chronic muscle aches
- Mood issues, including depression, bipolar disorder, and anxiety
- Immune dysfunction, such as autoimmunity or being virus-prone
- Detox problems, such as heavy-metal toxicity or yeast overgrowth
- Infertility
- Recurrent miscarriages
- Insomnia

remodeling (also known as nucleosome remodeling). Specific enzymes regulate gene expression by moving or restructuring nucleosomes. One class of enzymes that remodel chromatin are tumor suppressors. Women who have enough of these enzymes have a reduced risk of ovarian cancer.[13]

Your genes and lifestyle control your ability to methylate. It's a bit of the chicken-and-egg quandary. For example, consider the following scenarios I've seen in my functional medicine practice.

- You may have totally normal genes for methylation but eat so much sugar that you have dysbiosis, an imbalance of good and bad bacteria in your gut. The dysbiosis may block you from absorbing key B vitamins and cause you to methylate poorly.

- You may have faulty genes for methylation but eat so well and attend to your gut so assiduously that methylation is normal.

- You may have had a period of poor methylation, and, once it's addressed, you may become a hypermethylator. Seems like a good thing, but it makes you feel like crap, like you're coming down with the flu, because you are processing the backup of toxins that have been stored in your body. It's just temporary, but it can make you feel worse before you feel better.

While I can make general recommendations for most people to improve their methylation, a word of caution: some of you who are more complex or don't respond in the expected way to the Younger protocol may benefit from the one-on-one care of a functional medicine expert. To learn more about how to work with someone knowledgeable about the gene/environment interaction, go to functionalmedicine.org for a list of certified functional medicine practitioners in your area.

Sticky Notes on DNA

I explained in chapter 2 that gene regulations starts early with genetic imprinting in your mother (or grandmother). Jeffrey Bland, Ph.D., founder of the Institute for Functional Medicine and author of *Disease Delusion*, taught me that you can think of genetic imprinting as being like a sticky note applied to your chromosomes during certain key stages of your development. An example is the first trimester of life. When your mom was pregnant with you, if she survived a famine or got exposed to a particular toxin, a little sticky note was applied to your DNA, serving as a visual cue for your DNA to pay attention to that particular site at a later date.

Here's another example of genetic imprinting. During the great ice storm from Ontario to Nova Scotia in 1998, there was a vicious cold snap, and the severe subzero weather led to power outages and extreme cold exposure in people's homes for several weeks. Researchers tracked women who were pregnant at that time and their children to assess what type of genetic tags were placed by the severe prenatal stress. Their results were then compared with normal women.

The DNA methylation signatures were different in the children that were in utero during the ice storm and cold snap. Researchers examined the immune cells of these children and found profound differences in the methylation of specific regions of genes called the promoters, or the switches for genes to be turned on or off. The affected genes controlled insulin and blood sugar. It looks like these children, now fifteen to seventeen years old, will have a greater prevalence of diabetes in the next few decades. This is a powerful example of the *epigenetic switch* that can occur during severe stress.[14]

Some forms of genetic imprinting are more flexible than others, meaning that the sticky note can be readily taken on or off the chromosomes. Examples of behaviors that can affect these flexible imprints include eating nutrient-dense foods instead of trans-fats, walking briskly most days of the week, and not sitting at a desk for more than three hours each day. In short, even if you have faced severe stress or trauma or made poor lifestyle decisions earlier in life, you may be able to change the sticky note and thereby alter the way your genetic blueprint is read.

Even though we are early in our understanding of the interaction of a person's DNA with his or her exposome, proven solutions in the protocol start in the next chapter. Parts of it may be unquantifiable because of the number and complexity of influences; still, we are at a tipping point where we know enough to improve your exposome. Gene expression, as mediated by the exposome, happens continuously. Upgrade your lifestyle decisions as described in this book, and you will improve your environmental inputs and, as a result, your gene expression and health. Keep in mind that your daily choices likely have tremendous influence over your own health and the health of future generations. It's time for a change order to the blueprint of your body.

CHAPTER 4

Get to the Root

All parts of the body which have a function if used in moderation and exercised in labors in which each is accustomed, become thereby healthy, well developed and age more slowly; but if unused and left idle they become liable to disease, defective in growth and age quickly.

—Hippocrates

This morning, as I walked into my local barre fitness studio, I saw a woman with the body of a thirty-year-old, a face that looked about fifty, and a head of thick, gray hair. *Huh?* I couldn't stop staring. She was on a bicycle, which she handled with confidence, as if it were a favorite article of clothing, and then she casually locked it to a rack with the air of a seasoned cyclist. Bike helmet still on, the woman walked briskly into the studio ahead of me.

I know I'm not supposed to compare my body to someone else's, but I couldn't help myself; she was my height but leaner with better muscle definition and no "thass" (my kids tell me that you want to have a sharp demarcation between your thighs and ass; when they merge—giving you a thass—it's a telltale sign of aging). She headed to the front of the room (I prefer the back row). During class, I noticed that her handheld weights were heavier than mine, and her planks and push-ups were "full form" (barre lingo for on her hands and feet, not modified on the knees). She took every challenge thrown down by the instructor. Impressed and inspired, I had to investigate.

Sylvia is seventy-one. Her healthspan score is 74, above average. She's always been an athlete, although the rest of her family has little interest in fitness. Her resting heart rate is in the fifties. She is an omnivore who eats a lot of fruits and vegetables, and she eats almost exclusively at home. When she does go out, she orders a salad, a soup, or fish. She hates cars and doesn't own one, preferring to walk or cycle everywhere. Last weekend, she completed a metric-century ride (a bike ride of sixty-four miles—or a hundred kilometers—in four and a half hours). Before and after a long ride or workout, she sleeps nine hours per night. When she trains fewer than two hours in a given day, she sleeps a minimum of eight hours and naps, if time permits, for anywhere from fifteen to thirty minutes.

Sylvia is a publicist and former marketing executive, still working fulltime. In fact, she came up with the slogan "A woman's place is on top" for a T-shirt that was sold to fund the first women's ascent of Annapurna in the Himalayas in 1978. The intrepid female mountaineers sold more than ten thousand T-shirts, which enabled them to make the climb (Sylvia was invited to join them but had to decline). The slogan remains a fitting description of Sylvia and her personal style of aging slowly.

I explained to her that I was writing a book about aging, and she started peppering me with questions, her bright, alert eyes boring into me. She told me the secret to staying young was to remain open to new ideas and be curious. As Sylvia has done for several decades, you can age more slowly once you understand the $^{90}/_{10}$ rule about your genetic code. So let's prepare for the best epigenetics possible. I've gathered the most proven approaches that will help you age more slowly and stave off infirmity. We'll start by calculating your current healthspan score.

Take the Healthspan Quiz

In order to slow down aging, you need to measure your current aging process because what you measure improves. Additionally, it's important to start with a baseline. Considerable debate exists about the best way to measure aging, and there's not much consensus. Geneticist and Nobel laureate Elizabeth Blackburn says it's telomere length measured in white blood cells, a test done by specialized labs. Dr. Oz suggests his RealAge Test, a unique calculation of a body's health age, created by himself and Dr. Mike Roizen. They developed it back in 1999 and it needs to be updated based on the latest science. Many antiaging researchers rely on handgrip strength—your ability to squeeze your hand and bear weight, such as when you're hanging from a pull-up bar. Luigi Ferrucci, director of the Baltimore Longitudinal Study on Aging, uses handgrip strength to track aging and inflammation in a large cohort of healthy people he's been following since 1958.[1] Ferrucci also uses a blood marker that measures inflammation, interleukin-6 (IL-6), which your immune system makes; it's associated with 807 genes related to aging and mortality, and IL-6 levels rise as you get older.[2] Another system suggests that measuring the blood level of C-reactive protein, an indication of inflammation in your body, may be the best method for predicting mortality risk.[3] But none of these measurements is easy, convenient, and cutting edge. So I created my own, the Healthspan Score.

In order to take the quiz efficiently, you will need the following tools:

- Computer or tablet, if you prefer to perform the quiz online

- Measuring tape (in inches or centimeters) for recording your waist circumference

Option to take the quiz online: www.HealthspanScore.com.

Your Healthspan Score identifies the top priorities in several areas related to aging. And you'll begin to learn what to do about those. Along with doing the following quiz, I suggest that you take a before photo of your body and a close-up of your face (especially your eyes and skin) so you have a benchmark for your starting point. Here are the baseline measurements affecting your healthspan that you need to record prior to the test:

1. Resting heart rate

2. Waist circumference

3. Weight, height, and BMI

4. Fasting blood sugar (most mainstream doctors test this on everyone by age forty-five)

Healthspan Quiz

Take my healthspan quiz to determine the rate at which you are aging and set your baseline for calculating improvement. For each question, choose the answer that best describes you currently. Write down the number of points for each answer in the blank line provided. At the end, tally the subtotal for each section and the grand total (your Healthspan Score) in the space provided at the end of the quiz.

Demographics

1. Are you . . .

    female = 1 point
    male = 0 point   Subtotal _____

2. Age

    < 40 = 2 points
    40–65 = 1 point
    > 65 = 0 points   Subtotal _____

3. What's your waist size (measure waist circumference at belly button)?

    If female,   < 35 inches (< 88 cm) = 2 points
                  35 inches (88 cm) or greater = 0 points
    If male,   < 40 inches (< 102 cm) = 2 points
                  40 inches (102 cm) or greater = 0 points   Subtotal _____

4. Let's calculate your current BMI.

    Enter height in inches or cm: _____
    Enter weight in pounds or kg: _____

    Use an online calculator[4] or the formula **BMI = weight ÷ height^2**.
    If you used kilograms and centimeters, you have your result. If you
    used pounds and inches, you need to multiply the result by 703. For
    example, for a woman who is 150 pounds with a height of 64 inches,
    BMI = (150 ÷ 64^2) × 703 = 25.7

                                                       Calculate BMI_____

    BMI < 18.5 = 0 points
    BMI 18.5–24.9 = 2 points
    BMI 25.0–29.9 = 1 point
    BMI >30 = 0 points   Subtotal _____

                                       Demographics Total _____

Healthspan Quiz

Lifestyle

5. On an average workday, how many hours do you spend sitting?

   < 3 = 2 points
   3–6 = 1 point
   > 6 = 0 points   Subtotal _____

6. How much do you sleep on most nights?

   < 4 hours = 0 points
   5 to 7 hours = 1 point
   7–8.5 hours = 2 points
   > 8.5 hours = 1 point
   I'm not sure = 0 points   Subtotal _____

7. Do you exercise for at least 30 minutes, 5 days per week, at a moderate to vigorous pace?

   Yes = 2 points
   No = 0 points   Subtotal _____

8. Do you brush your teeth twice per day or more?

   Yes = 2 points
   No = 0 points   Subtotal _____

9. How often do you floss?

   Twice or more/day = 2 points
   Once/day = 1 point
   Less than once/day = 0 points   Subtotal _____

10. How often do you engage in some form of contemplative practice (yoga, meditation, mindfulness, tai chi, etc.)?

    5 times/week or more = 2 points
    1–4/week = 1 point
    Not at all = 0 points   Subtotal _____

11. How much alcohol do you consume each week?

    None = 0 points
    1–2 servings = 2 points
    3–7 servings = 1 point
    > 7 servings = 0 points   Subtotal _____

Healthspan Quiz

12. How many hours of sleep do you feel like you need to function well during the day?

    Less than 4 = 0 points
    5–6 = 1 point
    7–8.5 = 2 points
    More than 8.5 = 0 points
    I'm not sure = 0 points   Subtotal _____

13. Have you smoked at least 100 cigarettes in your life?

    Yes = 0 points
    No = 1 point
    I'm not sure = 0 points   Subtotal _____

                                                    Lifestyle Total _____

Health

14. Would you rate your physical health as better than others your age?

    Yes = 2 points
    No = 0 points   Subtotal _____

15. Do you wear sunscreen, avoid the sun, and/or have low vitamin D?

    Yes = 0 points
    No = 1 point   Subtotal _____

16. Have you been diagnosed with any of the following (0 points for every yes and 2 points for every no; 0 points if you don't know)?

    Diabetes or prediabetes
    Depression
    Alzheimer's disease
    Cancer (any type)
    Multiple sclerosis
    Gingivitis
    High blood pressure
    Heart disease
    Abnormal Pap test results (for females)
    Stroke
    Seasonal affect disorder (SAD) or winter blues   Subtotal _____

Healthspan Quiz

17. What is your resting heart rate (heart rate while sitting still)?

    Fewer than 60 beats per minute = 2 points
    70–79 beats per minute = 1 point
    80 or more beats per minute = 0 points
    I'm not sure = 0 points   Subtotal _____

18. Is your most recent fasting blood sugar between 70 and 85 mg/dL?

    Yes = 2 points
    No = 0 points
    I don't know = 0 points   Subtotal _____

19. Do you experience frequent colds or other types of infections (e.g., cold sores or herpes, respiratory infections, bronchitis, sinusitis)?

    Yes = 0 points
    No = 1 point   Subtotal _____

                                         Health Total _____

Skin, Hair, and Nails

20. Do you have weak, thin, or brittle nails?

    Yes = 0 points
    No = 1 point   Subtotal _____

21. Do you have white spots on your nails?

    Yes = 0 points
    No = 1 point   Subtotal _____

22. Do you have skin problems such as eczema, rashes, and/or acne?

    Yes = 0 points
    No = 1 point   Subtotal _____

23. Have you experienced hair loss?

    Yes = 0 points
    No = 1 point   Subtotal _____

                            Skin, Hair, and Nails Total _____

Healthspan Quiz

Stress

24. Have you experienced in the past 12 months a major life stressor such as the death of a loved one, divorce or separation, the loss of a job, or a move?

    Yes = 0 points
    No = 2 points   Subtotal _____

25. Do you feel that, more often than not, you're rushing from one task to the next, and you're stressed due to a lack of time?

    Yes = 0 points
    No = 2 points   Subtotal _____

26. Would you rate your life as very stressful?

    Yes = 0 points
    No = 2 points   Subtotal _____

27. Over the past two weeks, how would you rate your ability to handle stress?

    Poor = 0 points
    Moderate = 1 point
    Excellent = 2 points   Subtotal _____

    Stress Total _____

Food Intake

28. Do you eat foods with flour or sugar more than twice per week?

    Yes = 0 points
    No = 1 point   Subtotal _____

29. Do you eat at least seven servings of vegetables and fruits (1 serving = 1/2 cup) every day?

    Yes = 2 points
    No = 0 points   Subtotal _____

30. Do you eat at least one serving of greens each day (1 serving = ½ cup)?

    Yes = 2 points
    No = 0 points   Subtotal _____

Healthspan Quiz

31. Do you eat processed or packaged food, fast food, or food containing trans-fats (such as doughnuts, cookies, crackers) once per week or more?

    Yes = 0 points
    No = 1 point   Subtotal _____

                                          Food Intake Total _____

Family History

32. Do you have a family history of any of the following (0 points for every yes and 1 point for every no)?

    Alzheimer's disease   Diabetes
    Heart disease   Osteoporosis
    Stroke   Cancer   Subtotal _____

                                       Family History Total _____

Connectedness

33. Are you currently married or in a relationship with someone with whom you can share the experiences of daily life?

    Yes = 2 points
    No = 0 points   Subtotal _____

34. Do you feel isolated or lonely?

    Yes = 0 points
    No = 2 points   Subtotal _____

35. Do you feel enthusiastic and excited about what you're doing in your life?

    Yes = 1 point
    No = 0 points   Subtotal _____

36. Do you feel that there is someone in your life who cares about you and loves you no matter what?

    Yes = 1 point
    No = 0 points   Subtotal _____

Healthspan Quiz

37. Do you believe that you matter as an individual and that you make a difference in the lives of others?

    Yes = 1 point
    No = 0 points   Subtotal _____

                                    Connectedness Total _____

Oxidative Stress

38. Are you tired on a regular basis?

    Yes = 0 points
    No = 1 point   Subtotal _____

39. Do you experience fatigue after exercise?

    Yes = 0 points
    No = 1 point   Subtotal _____

40. Are you sensitive to smoke, perfume, cleaning supplies, or other chemicals?

    Yes = 0 points
    No = 1 point   Subtotal _____

41. Do you feel muscle or joint pain?

    Yes = 0 points
    No = 1 point   Subtotal _____

42. Do you smoke or are you exposed to secondhand smoke?

    Yes = 0 points
    No = 1 point   Subtotal _____

43. Are you exposed to environmental toxins, such as pollution, heavy metals, or chemicals, at home or work?

    Yes = 0 points
    No = 1 point   Subtotal _____

44. Do you take prescription or recreational drugs?

    Yes = 0 points
    No = 1 point   Subtotal _____

                                    Oxidative Stress Total _____

Healthspan Quiz

Brain Function

45. Do you struggle once or more per week to find the right word in a conversation?

    Yes = 0 points
    No = 2 points   Subtotal _____

46. Do you feel you've witnessed a decline in mental sharpness, memory, or focus in the past 5 or 10 years?

    Yes = 0 points
    No = 2 points   Subtotal _____

47. Do you feel like your brain is not functioning as well as it did 5 or 10 years ago?

    Yes = 0 points
    No = 2 points   Subtotal _____

48. Do you have impaired taste, smell, and/or hearing?

    Yes = 0 points
    No = 2 points   Subtotal _____

Brain Function Total _____

49. Do you believe that chocolate, wine, and guacamole help you look and feel young?

    Guess what—they can! (1 point for yes, because you have a sense of humor.)

Grand Total _____ **/100**

Date _____

Interpretation

You now have a baseline measurement of the most important factors determining your rate of aging: demographics, lifestyle, stress, exposures, medical and family history, antioxidant status, connectedness, and brain function. What's next? Use the table below to interpret your score. You will calculate your score again after the seven-week protocol and then periodically in the future (I suggest every six months) to make sure you're on track to lengthen your healthspan. The rest of this chapter will help you get ready to take back control of the aging process and your health. In the future, if your score ever drops, look closely at the areas where your score decreased and then review the correlating protocol/rituals in this book.

Score	What it means
< 40	Very Low Healthspan. You are aging excessively. In addition to starting the Younger protocol, see your doctor for help and accountability while you slow down your aging.
40–49	Low Healthspan. You are aging fast and are at high risk of a compromised healthspan. You don't have time to waste. Start the Younger protocol as soon as possible.
50–59	Below-Average Healthspan. You are aging moderately fast, but this protocol will help you age more slowly.
60–69	Average Healthspan. Your healthspan is in the average range, but we have a lot of work to do together.
70–79	Above-Average Healthspan. You are doing a lot right, but we need to address several gaps.
> 80	Excellent Healthspan. Perform the Younger protocol to reinforce the good practices you now have in place and establish them as habits, looking for the small tweaks that will improve your score.

Why These Measurements Matter

These measurements indicate how your genes are performing. Every one of them affects your genetic expression and helps highlight, even *prioritize*, the functional medicine solution. Each section reflects a key aspect of aging, from disease risk to oxidative stress, and therein suggests where you need the most help. For example, if you did well in most sections except lifestyle, then that's where to give more focus in the seven-week protocol. Each chapter of the Younger protocol will improve your healthspan, even in areas where your score is already on track. If you eat an antiaging food plan but skimp on sleep, then we need to address that. If you've honed your lifestyle but you have relatives with Alzheimer's disease, then chapter 11 ("Think") will be more important for you.

One measurement alone doesn't necessarily tell the whole story. As an example, let's look at heart rate. My husband has a resting heart rate of forty-nine because he's an athlete who excels at every sport. My resting heart rate is around sixty, on a good day. That means he is more efficient at pumping blood from his heart to the rest of his body than I am. It's part of the muscle factor. The key to slowing down aging is slowing down your resting heart rate. Endurance-oriented physical exercise lowers your resting heart rate and reduces your total heartbeats over twenty-four hours. Additionally, endurance athletes have better balance in their nervous systems than nonathletes. They rest and digest better but can kick it into high gear and perform when needed.

The health improvements in athletes are not confined to heart rate; they extend to beat-to-beat variability, known as *heart rate variability* (HRV). Your HRV is the pattern of your heart rate—if you have a resting heart rate of sixty, you might think that's one beat per second, but that would be very low HRV. You actually want variability between the time of each heartbeat; for example, the first gap between heartbeats is 1.00 seconds, and the second is 1.02 seconds, and the next 1.05 seconds, and so on, in order to indicate good HRV.

I think of HRV as a measure of the flexibility of your nervous system, and more flexibility is better. A certain level of variability is considered healthy and advisable. The variability results from several factors, including external (lifestyle, behavioral, and environmental) and internal (neural reflex, neural central, hormonal and other humoral influences).[5] Think of a low resting heart rate and increased HRV as being associated with longevity.[6] (You'll learn more about how to measure HRV in

Test Your Vision

It's common but not unavoidable for people over forty to need glasses. Usually the problem is presbyopia, which is when you experience blurred vision during near work, such as reading, looking at your cell phone, sewing, knitting, or working at the computer. Even if you've previously had other types of eye problems—nearsightedness, as an example—you may find you have blurred vision even when you wear your usual glasses or contact lenses. You may need to hold the object you're reading at arm's length in order to see it more clearly.

Presbyopia results from aging, unlike nearsightedness, farsightedness, and astigmatism, which tend to occur in childhood or young adulthood. The most likely explanation for it seems to be that proteins in the natural lens of the eye begin to age, leading to gradual thickening, hardening, and stiffness. Additionally, the muscle fibers surrounding the lens get older, and what you notice is that you have more difficulty focusing on an object up close.

Presbyopia rates are rising, mostly because the population is growing older. Furthermore, the amount of near work people do has risen; just look at the exponential uptick in the use of cell phones, handhelds, and laptops.

Test your vision by downloading an eye chart from the Internet and checking each eye from a specific distance, just like an eye professional. See the Resource section at the end of the book for apps and other ways to measure your vision easily.

What can you do about it? I'll give you specific exercises in chapter 9, but before beginning the Younger protocol, take a digital detox once a week for twenty-four hours, meaning you ban cell phones, computers, TV, and handheld devices. For every forty-five minutes of near work, take a fifteen-minute break. More daylight when performing near work may help. Plus, spend more time outdoors, where you have to focus on seeing objects at greater distances.

chapter 7.) However, HRV is just one measurement. An athlete could have a high HRV and yet sleep too little, have a family history of Alzheimer's, and work full-time in a factory where she is exposed to toxins forty hours each week. That's why your Healthspan Score is an aggregate measure that doesn't rely on a single factor to reflect your rate of aging.

Help Wanted: New Paradigm

People in their sixties, seventies, and eighties often ask me if it's too late to age gracefully, but what cutting-edge age researchers know about the innate intelligence of the body is that it keeps flexing, adjusting, and adapting to the right cues until the day a person dies. So it's never too late to deactivate the time bombs in your cells and boost your healthspan.

Regrettably, we live in a world that is at odds with our DNA. There's a mismatch between genes, life span, and our dominant culture—the societal norms that guide educational, marital, work, and retirement choices. Currently, societal norms assume that you exponentially decline starting around age fifty and then retire ten to twenty years later when you are no longer productive. Yet our genes offer a distinct alternative. Maximize what your genes can do by improving your environment, regardless of age, as Sylvia in my barre class has done.

To accomplish this, we need a new paradigm about aging so that we live longer, live better, and feel more vital. A paradigm shift requires that you reconsider many aspects of aging, from failing vision to how much time you sit during the day to your life purpose. Why not prepare yourself for a better quality of life for a longer amount of time? Start thinking about the benefits of being able to pluck out worn-out cells as if they were rogue chin hairs. That's the benefit of implementing the Younger protocol.

Keeping Up with the Icarians

Let's get ambitious so we can match the achievements of long-lived residents in places such as Monaco, Icaria, and Okinawa. Not only do they live longer and with great health to the end, but they also have a far better quality of life than we do in the United States. Consider the people of Icaria (sometimes spelled Ikaria), a mountainous island in Greece. I suspect they will outlive the Kardashians!

Icarians live ten years longer than most Europeans; they run up and down their hilly landscape daily, at all ages and in much greater health. Named after a figure in Greek mythology, Icarus, who died at an early age because his artificial wings fell apart when he flew too close to the sun, the island ironically boasts the most ninety-year-olds in the world, with one in three people there surviving to ninety. It has been dubbed "the island where people forget to die,"[7] but I prefer the nickname "the island where people don't lose their minds," as the population is almost completely devoid of dementia and depression. Icaria is more isolated than other Greek islands (it's about a ten-hour ferry ride from Athens), so it's been spared most of the trappings of tourism, including fast food and a fast-paced life. As a result, the island even now is a great laboratory for a different way of life, one that happens to be associated with a long healthspan.

While no single factor explains longevity across the board, it's fun to peek inside the typical day of an Icarian to see how he or she achieves such a long healthspan.[8] Most of the data collected on Icarians has been gathered by the University of Athens, the Harvard School of Public Health, and journalist Dan Buettner in his investigations for *National Geographic* to study the globe's longest-lived cultures.[9] Consider these factors as we start to cover the steps of your own Younger protocol.

- Wake up naturally, without an alarm clock, and don't put on a watch. Icarians don't wear watches and have a relaxed attitude about time.

- Bathe in curative hot springs. Hippocrates, the father of modern medicine, considered hot mineral baths curative. And in Europe and Japan, doctors widely accept them as a form of therapy for knee pain, arthritis, fibromyalgia, high blood pressure, eczema, and other problems. Natural hot springs include various minerals such as sulfur, thought to improve nasal congestion; calcium and sodium bicarbonate, believed to enhance circulation; and salt, presumed to help digestion.

- Eat lots of fish, greens, and other fresh vegetables. Even compared with the standard Mediterranean diet, Icarians eat more fish and fresh vegetables, especially wild greens such as dandelion, fennel, and horta (a cousin of spinach)—more than a hundred and fifty varieties of local greens grow wild. They rarely eat meat, usually once a week or less, and liberally drizzle olive oil as a condiment on food at the dinner table. They eat six times more beans than Americans and a quarter of the sugar. Most people have access to a family garden and livestock such as goats. But locals stress that it's not the food alone; it's enjoying the food in combination with conversation with loved ones.

- Know your neighbors and socialize often with friends and family. Strong social connections improve health and longevity. Icarians are famous for their open-door lifestyle and broad invitations to visitors to join them for a slow, friendly meal.

- Consume raw, unpasteurized goat's milk. Icarians consume unpasteurized goat's milk and use it to make yogurt and cheese. It's known to be hypoallergenic compared with cow's milk and does not bother most people with lactose intolerance. While goat's milk is seemingly healthier than cow's milk for you, it's the raw part that may matter most for your health. When milk is pasteurized, it kills the probiotic *Lactobacillus acidophilus*. You need *Lactobacillus acidophilus* to make B vitamins and inoculate the gut with healthy bacteria.

- Walk like a goatherd, and garden. The rugged mountain terrain of Icaria requires a mini-workout each time someone leaves home. Sixty percent of Icarians over ninety are physically active, compared with about 20 percent elsewhere. According to visitors, it's hard to get through the day without hiking at least twenty hills.

- Drink wine moderately. Locals explain that their wine is pure, nothing added, no preservatives. They drink two to four glasses per day. When consumed with plenty of fruits and vegetables, wine nudges the body to absorb more flavonoids, a type of plant derivative shown to benefit your health.

- Fast intermittently. Most Icarians are Greek Orthodox Christians, and their religious calendar calls for intermittent fasting about six months of the year. Prior to an Orthodox feast day, they don't eat for eighteen hours. Occasionally restricting food has been shown to slow the aging process in mammals.

- Nap each afternoon. Icarians' standard practice is to nap after lunch for thirty minutes, at least three times a week but sometimes daily. Did you know that napping lowers your risk of heart disease by 37 percent? I didn't know that either, but the mechanism seems to be related to lowering stress hormones and resting the heart.

- Eschew retirement. Icarians have a relaxed attitude about getting to work in the morning, but their work gives their lives purpose and meaning. They do not believe in retirement and view work as a way of life, not something separate from it. To them, it's all sacred time.

- Drink a thick mountain herbal tea brewed from marjoram, spleenwort, purple sage, rosemary, oregano, chamomile, dandelion leaves, artemisia, or a wild mint called *fliskouni*. Many Icarian herbal teas act as diuretics, which flush waste out of the body and lower blood pressure by removing excess sodium and fluids. Icarians enjoy their mountain tea as a tonic at the close of day.

Functional Medicine: Stop Sitting on the Tack

Before we improve your healthspan by making room in your life for the epigenetics that will keep you healthy well into old age, just like the people of the Greek island of Icaria, let's address the key problems that steal your youth. In standard medicine, the convention is to apply prescriptions to

Ten Conditions That Steal Your Youth

I'm obsessed with what makes you old before your time—the conditions that cause *inflammaging*. How do you know if you have inflammaging? You feel stiff, slow, tired, and can't remember why you walked into a room. Here's a quick list of the most common problems that contribute to inflammaging and shorten your healthspan.

1. Getting fat

2. Sitting too much

3. Taking certain medications, like antianxiety pills or even antihistamines (like Benadryl)*

4. Eating too many carbohydrates and processed foods

5. Losing muscle (no longer building strength and regularly using muscle fibers)

6. Sleeping less

7. Lacking vision and purpose

8. Getting insufficient vitamin D

9. Feeling stressed out

10. Isolating socially

*Antianxiety pills such as Valium, Xanax, and Ativan were shown in a recent study to raise the risk of Alzheimer's disease by more than 50 percent. Further, people who take Benadryl for sleep or allergies should reconsider, if they can remember—the *Journal of the American Medical Association* just published a study linking frequent and long-term use of anticholinergic drugs like Benadryl with dementia.

address symptoms rather than to treat the root cause. The problem with this approach is that you may feel better temporarily, but your medical condition keeps progressing and you age faster. Sydney Baker, M.D., one of the early practitioners of functional medicine, used to say that if you're sitting on a tack, the solution is to find and remove the tack, not to treat the pain. In functional medicine, the goal is root-cause analysis so that you tune up the biology, reverse the medical condition, and feel better in the long term. In fact, several key functional medicine recommendations will benefit nearly everyone and decelerate aging; those recommendations are part of the protocol that you will begin to learn about in this chapter.

Since the Younger protocol is based in functional medicine, the protocol addresses the ten most common root causes of accelerated aging by focusing on multiple genes as well as their interactions with one another and with your lifestyle, including your diet, hormones, toxins, stress, omega-3/omega-6 balance, vitamins and minerals, allergens, sleep, and exercise. Lifestyle factors affect the accumulation of free radicals and other damaging molecules, mitochondrial dysfunction, hormonal decline, telomere damage, and, finally, inflammation, or *inflammaging* (that unfortunate combination of progressive inflammation and heightened stress response that leads to faster aging).

The Younger Protocol Prerequisites

Before moving onto the details for the first week of the protocol, you need to meet the minimum three essentials of a long healthspan, or the prerequisites:

1. Sleep a minimum of six hours per night. If you don't meet this prerequisite, start spending an extra thirty minutes in bed. Sleep cleans out the garbage of your body. It's like a power cleanse. In chapter 6, I'll show you how to optimize your production of melatonin, an important antiaging hormone that controls more than five hundred genes.

2. Avoid processed food. If it didn't come out of or walk on the ground and isn't recognizable as a plant or meat or fish, avoid it. Of course, processed foods fall along a continuum: macadamia nuts are less processed than macadamia oil. The point is to avoid foods with five or more ingredients that you cannot easily pronounce, or fake food that comes in a box with a long shelf life. A vegetable bowl with chicken is okay, or flax crackers that contain simply organic flax

seeds, apple cider vinegar, sea salt, and herbs. A jar of pasta sauce with added sugar is not.

3. Exercise twenty to thirty minutes four days a week. Yes, that includes walking! Ideally, start to notice your heart rate at rest and while exercising, before refining your approach in chapter 7.

These three prerequisites need to be in place prior to week 1. Your body will not be able to benefit from the other aspects of the protocol until you establish these foundations, even if it takes you a few days to a few weeks to make them a habit. For people who are already performing these basic foundational steps, the prep phase should require only a day or two to get the supplies you need for success in week 1.

Define Your *Why*

Next, you need to articulate your *why*. You have a belief about aging that I want to unearth. This belief is your motivation to slow down aging. It's a touchstone that's probably what made you buy this book and it can be cultivated while you're developing new habits that will keep you young. It's your motivation to act, even when it's difficult or inconvenient. Your *why* is far stronger than willpower or following the protocol because you think it's a good idea. Your *why* is deeply personal and will sustain you over the long term.

My *why* for de-aging is that I want to live a long life with my husband, hiking our favorite trails in Point Reyes National Seashore, having long conversations that I cherish, watching our two daughters grow up into fabulous, interesting women, and taking care of our future grandchildren should the girls choose to have kids.

My husband's *why* is different, just as your *why* may not resemble mine. At his current age of fifty-six, he wants to augment his healthspan so that he can always be highly physical. David wants to be vital so he can still fish as he gets older and be able to turn his neck like a normal person, rather than like the former football fullback he is. He thinks it's a shame that so many people build financial equity over the decades leading up to retirement, but when retirement age comes, they lack health and are physically bankrupt. They have no health equity. He wants health equity. He needs to reduce the tightness and inflammation that he feels in his back, which drives him to need a chiropractic adjustment several times per month (and probably relates to his twelve years of playing tackle football).

He wishes to wake up in the morning feeling rested and happy to be alive, not in pain. He wants to eat more lemon-ricotta pancakes and to drink the most amazing gluten-free India pale ale (the one that hasn't been invented yet. *Damn!*). He wants to dance with our daughters at their weddings. He can't imagine he'll be alive to help me take care of grandkids, but if so, he's game. He's less optimistic than I am about aging, not to mention six years older. Nevertheless, his Healthspan Score is quite high, especially for his age.

My friend Jo Ilfeld is forty-two. She's married with school-age kids like me; we met in a moms' group. She describes her *why* in this way: "I want to see how my kids turn out. And I want to retire and do all those things I don't have time to do now. Or if I don't retire, I at least want a marriage without kids for a while. Oh, and I want to have great sex until I die!"

Prep for Your Protocol

When I first started writing books, my friend Jo asked me to offer two tracks: one that gives the minimum actions to achieve results and one that gives all the advanced stuff. If you're a skier, it's like a green run versus a double-black-diamond track. I'm giving you both. Here are a few tips.

- If you want to keep it simple, just perform the actions marked *basic rituals*.

- If you just want to know what to do and don't want to get lost in the science, skip the science sections in chapters 5 through 11 and go straight to the protocol sections.

- When you perform the protocol again in the future (I recommend twice a year), add an *advanced project*.

- If you want a greater challenge, perform the basic rituals plus the advanced projects.

- All the prep emphasizes the basic track, with a few notes for those with the time and energy to follow a more advanced track. Ready to get started? Me too.

Before Week 1: Upgrade Your Body with Food

- Complete Healthspan Quiz

- Go shopping

- Buy or make fermented food [10]

  - Sauerkraut

  - Cultured vegetables (my favorites are beets, turnips, and cabbage)

  - Kimchi

  - Coconut kefir

- Stock up on healthy fat

  - Coconut oil, preferably unrefined and expeller-pressed

  - Medium-chain triglyceride (MCT) oil, a very efficient type of oil derived from coconuts that gets rapidly converted into energy for your brain and body because it doesn't require a stop at the liver for processing. (You don't need bile acids to digest it, so it's easier on your GI tract.)

  - Grass-fed butter (made from cows that graze on a grass pasture) or ghee (clarified butter)

  - Chia seeds

  - Flaxseeds

- Avocados

- Marine fats such as an omega-3 supplement, wild-caught fish (salmon, cod, halibut), krill oil

• Purchase clean protein and have about three to four ounces of animal or plant protein at each meal, which will help switch on your longevity genes. Ideally, eat meats only from animals fed in their natural habitats: pastured chicken and grass-fed beef, buffalo, and venison. Limit pork and processed meats such as sausage.

• Set your goal of low and slow carbs to decrease inflammation and glycation. Stock up on sweet potatoes, yams, yucca, and quinoa.

• Make or buy bone broth. It's rich in collagen, a protein needed for skin, teeth, and nail health. Your body's production of collagen declines with age, leading to wrinkles, neck waddles, and weak joint cartilage. For our family, making bone broth is the most convenient way to get collagen into our food plan. If that sounds disgusting, just start with chicken bones, filtered water, and a slow cooker—the slow cooking breaks the collagen down into gelatin. You'll be amazed. See the Recipe section at the end of the book for fish, chicken, and beef bone broth.

• Stock up on organic red wine if you drink alcohol. Yes, it's better than white wine, beer, or cocktails (more on this topic in chapter 5).[11]

• Pick up a bottle of berberine. Blood sugar rises with age, starting at fifty, and berberine is one of the supplements proven to help you normalize serum glucose.[12] Not only that, berberine will cool inflammation in your body, lower cholesterol, assist weight loss, and behaves like an antioxidant.[13] I recommend it if your fasting blood sugar is greater than 85 mg/dL. Take 300 to 500 milligrams once to three times per day, which has been shown to activate an important enzyme called adenosine monophosphate-activated protein kinase, or AMP, nicknamed "metabolic master switch." Talk to your pharmacist if you take medications (such as certain antibiotics) to make sure it doesn't interfere with drug metabolism. Beginning in day 1, take berberine every day, before or during a meal and throughout the protocol. Add in milk thistle to increase efficacy. Stop after two months to allow liver enzymes to normalize.[14]

In the first week of the protocol, you will learn the counterintuitive but easy-to-follow rules related to food, drinks, oral health, and supplements that will defuse the time bombs in your cells. The focus will be on the actions you can take to create "magic wands" that enable the body to produce enzymes, hormones, and other substances essential for slowing down the aging process. Recommended resources to help you along the way are in the appendix at the end of the book. Ready to learn the daily choices that help you defy genetic tendencies and fight disease?

CHAPTER 5

Feed

WEEK 1

Human suffering and misery comes from diseases that
should have been preventable but were not.

—Francis Collins, director of the National Institutes of Health

B ig birthdays: 40, 50, 65. Customarily, they prompt you to catapult
your health or finally ditch a bad habit. Life seems shorter, more
fleeting. Suddenly you want a bucket list. No more bikinis or
Speedos. You wonder about how your life is progressing (or not), and how
to curate the remaining years. You want to deconstitute the old you and
reconstitute the new and improved version. When I turned fifty this year,
I became obsessed with ways to lengthen middle age—and extend the
period when you feel awesome. You don't need a big birthday to receive
the wake-up call. Many people are now going into their mature years (defined as sixty-seven and older) feeling full of life, and that's what I want for
you. Possible? Yes, with a few caveats and starting first with your mouth.

In week 1 of the Younger protocol, you will learn how habits related to
your mouth, including eating, drinking, flossing, brushing, and supplementing, can alter your genetic expression. It starts with food, because
food is not just fuel—it's actionable information for your DNA, one bite
at a time.

Which genes are being altered? you might ask. We will be *turning on* the longevity gene for sirtuin (SIRT1), which protects you from aging-related diseases by revving up the mitochondria, the power plants inside cells that tends to weaken as you get older. You'll learn about intermittent fasting, which turns off an anti-longevity gene: mechanistic target of rapamycin (mTOR). We'll activate the vitamin D gene (VDR) and the seafood gene (PPARγ). We will be *turning off* the gene associated with Alzheimer's and heart disease (APOE4).

Why It Matters

Food doesn't do you in. Life does, by the myriad ways it stresses you out, accelerates your weight gain, muscle loss, cognitive decline, overzealous immune attacks on your own tissues (known as autoimmunity), and, generally, inflammaging. Some liken inflammaging to a faulty thermostat that makes the level of inflammation in your body either too high or too low, leading to damage and aging cells—and causing, respectively autoimmunity or cancer. Inflammaging is the default pattern. In fact, it's easier to shorten your life than it is to extend it.

The fastest ways to age are to gain weight, screw up your blood sugar, economize on sleep, sit a lot staring at a computer screen, feel chronically stressed and anxious, and eat the top inflammatory foods: sugar, gluten, and dairy. So it makes sense that in week 1 of the Younger protocol, we will start with addressing the way you eat and drink.

If you struggle with your weight like I have, you know it's ridiculously easy to gain weight as you age. There's no need to binge or try at all—it just happens. Besides the five aging factors from the introduction (page 7), here's why it's so easy to get fat after thirty-five. It starts with a silent war waged between fat and muscle.

- After thirty-five, your body fat rises 1 percent per year unless you take specific action to build more muscle and fight the war.

- After forty, you lose muscle mass gradually. By age fifty, you've lost, on average, 15 percent of your lean body mass. By seventy,

Meet Betty

It was 2008 when I first saw Betty Fussell profiled in the pages of *Vogue* as a woman of accomplishment in her eighties. Long hair billowing behind her and holding a baby goat in her arms, Betty leaped off the page like an alluring force of nature, inspiring the awe I used to feel when I saw Mount Rainier on a rare clear day in Seattle.

Betty is eighty-nine years young and the author of eleven books that range from food history to cookbooks to memoir. She's taught at Rutgers, Columbia, and New York University, among other institutions. She is a wise and articulate advocate for badassery and a sensual life.[1] Betty believes most people need greater intimacy with their food, that they need to be involved in the daily process of what they put in their mouths. Quick example: Recently, Betty went hunting in Montana with her son and shot her first deer. She turned the hide into a deerskin blanket for her bed and delighted in a freezer full of venison steaks, sausage, and jerky for the winter.[2]

Seven years after I saw Betty Fussell in *Vogue*, I went to my friend Meryl's wedding-rehearsal dinner, and as soon as I opened the door to the Italian restaurant on the Upper West Side, I felt the warm wash of a hearty laugh emanating from the corner—Betty Fussell. Meryl welcomed me, kissed my cheek, and whispered excitedly: "I sat you next to my dear friend Betty Fussell, the great food writer!" She pointed to the table where Betty was holding court, her long hair piled up like a prima ballerina's, her eyes flashing. She was beautifully clothed and exuded the type of kinetic energy that made people want to be near her.

I sat down and introduced myself. As I explained that I was Meryl's roommate from medical school, Betty rolled her eyes. Not sure what her eye roll meant, I asked, and I listened as she told me that most doctors were too mechanistic and reductionist; that *doctors had been ruined by conventional medical indoctrination*. Then she offered me the breadbasket and the butter. I declined, and Betty saw this as evidence of my failings as a doctor. "It's not food that does us in," she said. "It's life." She had a point. Food has been falsely accused. The problem is the food we choose and our relationship to it. Even more broadly, we have a problem with a lot of things we do and don't put in our mouths.

it's 30 percent loss per decade! There's a name for age-related loss of muscle: sarcopenia. Some of the decline is due to the loss of testosterone, the hormone that builds muscle and stimulates growth and repair. Some of the muscle loss is due to another hormone that belongs to the family of growth factors, myostatin, which is a powerful negative regulator of the size of your skeletal muscles, so you want to keep it low. It turns out that myostatin may control loss of muscle mass in aging women, although the full story is still unfolding.[3]

- You lose fast-twitch muscle fibers first, the ones that put a spring in your step and allow you to jump and sprint. The fast-twitch fibers wane before you start to lose your aerobic capacity.

- Old fat ages badly. Think of butter or lard sitting out on your kitchen counter for a few months. Yuck! In other words, body fat is not inert. Aging makes you fat, and then your fat makes you age— it's a vicious cycle that's hard to break.

- Fat is bossy and tells your brain to eat more by making you deaf to the signals of insulin and leptin.

Not all fat is bad; brown fat in your neck and back keeps your body warm and your metabolism high. But white fat, particularly in your belly, where it's called visceral fat, invades your inner organs, injecting them with inflammatory messengers like interleukin-6 and TNF-alpha. These cause the low-grade burn that makes you wrinkled and stiff.

This news can be quite depressing; I understand. So let's change your epigenetics so that you can stay lean with taut skin and energy to chase your grandkids or travel the world (or both!).

Your Genes Have Little to Do with Your Weight Gain

As much as I'd like to blame my thick legs and tendency to gain weight on my peasant ancestors, when I started testing my own genes, I was surprised to find *only 3 percent of my weight is controlled by my genes.* That means 97 percent of my weight is due to my lifestyle—the food I eat, the beverages I choose to drink, the stress I encounter and clear (or don't), my hormone balance, my mind-set, my sleep quality, and how I exercise. This calculation is based on a study of a consortium of a hundred thousand adults of European descent; it found eight main genes that are most significantly linked to differences in BMI (a ratio of weight to height).[4] The important takeaway is that weight has far more to do with lifestyle than genes. About 80 percent of weight is directly or indirectly linked to the food we put in our mouths, and the remaining 20 percent is due to factors such as exercise, sleep, stress, hormones, microbiome, and genetics.

Food Intimacy

Betty Fussell certainly seems to have found what works. I asked what she felt was important when it came to creating a good life. She resisted my desire to formulate a protocol. (Betty and I disagree on this point. I get asked daily for practical steps and guidance, information that I feel is important to share.) Nevertheless, I listened intently to what she had to say: "Unless you're involved in the daily process of what you put in your mouth, you're not engaged with that as your relationship to the Other, which you are consuming, eating, putting inside you." She explained that until food becomes an intimate relation, you are lost. "You're liable to either gobble it down, discard it, or demonize it. You gobble it down as fuel with its fat content, caloric content, chemical content *blah blah blah*, never seeing the food for what it is, as a part of you."

She told me that her day was not complete unless she had become involved with food: assembling, touching, smelling, and enjoying it. Unlike many people, she's never seen cooking as a chore. She doesn't want to sit at a table and have food served to her—it's not intimate.

Betty lives in Santa Barbara, close to the downtown farmers' market, where she recently bought a heritage red tomato. "He's the reddest red tomato but he's getting a little squishy, so it's time to make him into a gazpacho. You need to know him, the tomato. You know the value of his best parts, and you can't help but honor him because you're looking forward to how he's going to go with that cute little fresh cucumber that also came from the market, the fresh raw garlic, the fresh raw ginger. You're imagining all that. You get to put it all together. It's like directing a play. You're so grateful for all the participants. And they all have to work together or else it doesn't taste good."

I pointed out that it sounded like she was planning to seduce her tomato. "Of course! It's like making love; it's all erotically based. It's sensual. Show me how a man eats, and I'll tell you how he makes love. The guy who gobbles his food down . . . no thanks." Betty's intimacy with food spans decades, and now it's time for you to become intimate with food, the kind of food that's going to extend your healthspan and enjoyment of all of life.

Science of Week 1: Feed

Skip this section and head straight to the protocol unless you're interested in the science. I find that knowledge motivates my behavior changes, but that's not true for everyone.

Thousands of peer-reviewed studies prove there's a link between what you put in your mouth and how many feel-good years you have. You can eat, drink, and floss in a way that alters your genetic expression and keeps your cellular energy high and your immune system bulletproof, but over the years I've found that most people are more interested in results than science. If that's the case for you, skip ahead to the Protocol for Week 1: Feed.

Nutrigenomics: Nutrient/Gene Interactions

Nutrigenomics refers to the ways that foods such as broccoli and matcha help you stave off physical and mental decline. It's the powerful new science of how *nutrients in your food, drinks, and supplements* may affect your health by *changing the expression of your genes.* This is the realm of "personalized lifestyle medicine" and the future of medicine—dietary intervention based on your nutrient status, nutritional requirements, and genes. This knowledge can be applied to prevent or cure diseases such as cancer and autoimmune disorders.

There's a particular approach for de-aging that works best: eat mostly plant-based food, with animal-based food as a condiment, and choose anti-inflammatory forms of protein and dairy. I've already asked you as a prerequisite (in chapter 4) to give up processed foods—that includes anything that doesn't grow in or run around on the ground. Eat foods that don't raise your blood sugar excessively, because foods that elevate your blood sugar give you only a quick rush followed by a crash. Ironically, you crave sugar and processed foods when you're burned out, but they are poor choices. Don't fret about the details of glycemic index or glycemic load, since they've not been proven to aid in weight loss, blood-sugar stability, cognitive function, or even athletic performance—at best, the results from large, long-term studies are inconclusive.[5] Instead, simply avoid processed food and foods that are excessively high in carbs, like French fries, ice cream, and chocolate cake. Eat real carbs like squash, quinoa, and sweet potatoes.

Now here's a look at what you *do* want to focus on eating.

Goal: Eat more vegetables, about 1 to 2 pounds or 5 to 10 cups per day. Minimize refined carbohydrates. Avoid sugar and processed foods.

> **Protocol:** Aim for a plate that is 80 percent vegetables and 20 percent protein. Omit refined carbohydrates. Eat limited real carbohydrates: starchy vegetables like squash and root vegetables, nuts, seeds, and tubers.

> **Scientific Rationale:** Reduce blood sugar and insulin resistance to turn off the genes related to fasting glucose, such as G6PC2, and the genes related to insulin secretion, such as TCF7L2. Preserve function of the beta cells of the pancreas, which produce insulin, if you tend to have impaired function (gene SLC30A8). Restore mitochondrial function.

Goal: Reduce unnecessary inflammation by avoiding foods most likely to cause intolerance.

> **Protocol:** Avoid gluten and dairy; minimize grains (eat none if you need to lose weight or have an autoimmune condition).

Scientific Rationale: Turn off IL-6, TNF-alpha, and CRP genes, which contribute to low-grade inflammation.

Goal: Get omega-3s.

Protocol: Eat wild-caught fish once or twice per week (3 to 4 ounces per serving for women; 6 ounces for men). Avoid seafood that contains mercury.

Scientific Rationale: Turn on PPARγ and vitamin D genes.

Goal: Consume more medium-chain triglycerides (MCTs) and avoid toxic fats such as trans-fats, corn oil, cottonseed oil.

Protocol: Use coconut oil for cooking; MCT oil and olive oil for salad dressings and for drizzling on steamed vegetables.

Scientific Rationale: MCTs help you feel more full than long-chain fatty acids found in vegetable oils and can help regulate the Fatso gene. Additionally, MCTs can turn off the Alzheimer's and Bad Heart gene.

Goal: Lose weight and mimic calorie restriction.

Protocol: Fast intermittently for 12 to 18 hours once or twice per week. For instance, stop eating at 6 P.M., and eat again at noon the next day for an 18-hour fast, which seems to work best in women. Be sure to eat a nutrient-dense meal before the fast to make sure you have the nutrients you need.

Scientific Rationale: Turn on SIRT1 and turn off mTOR longevity genes; boost autophagy.

Goal: Prevent diabetes and obesity.

Protocol: Eat homemade meals, 11 to 14 lunches and dinners per week.

Scientific Rationale: When middle-aged people eat more homemade meals, they lower their risk of diabetes by 13 percent and obesity by 15 percent.

Goal: Minimize toxic red meat and fat.

> **Protocol:** Avoid processed meat like sausage, hot dogs, deli meats, and bacon. Limit grass-fed red meat to 18 ounces or less per week. Grass-fed beef has higher levels of omega-3s than grain-fed beef.
>
> **Scientific Rationale:** Red meat is associated with a greater risk of heart disease and cancer in men and women. Processed meat is linked to heart disease and diabetes. Data on grass-fed meat are limited.

Nutrigenomics: Drink/Gene Interactions

You might not think much about what you drink and why, but some beverages slow down aging and some accelerate it. To start, avoid drinks that contain sugar, artificial sweeteners, and high caffeine. No more diet soda or juice. No more Red Bull. They can damage your mitochondria.

My body has a gene that codes for a version of an enzyme that makes me break down caffeine very slowly, so a cup of coffee in the morning makes me stressed, jittery, and bitchy. Plus I won't sleep well that night. In general, the effects of drugs like alcohol and caffeine are greater if you are genetically programmed to break them down more slowly. So I had to discover the circumstances under which I could drink it, if there were any. People like me have a greater risk of heart disease when they drink coffee, whereas those who metabolize caffeine quickly receive a longevity benefit. The same may be true for alcohol.

The key is to decipher whether caffeine is aging you. To find out, do this simple experiment. Switch this week from coffee to a morning beverage that is lower in caffeine by at least half, such as green tea, white tea, yerba mate, or guayusa, a "clean energy" herb from the Amazon. (See chart for caffeine content of drinks.) Notice what happens to your energy over the course of the day and how well you sleep. If you feel less jacked up and your sleep improves, you may be a slow metabolizer of caffeine, like me.

CAFFEINE CONTENT
OF DRINKS

COFFEE, 1 CUP .... caffeine: 100 mg

GREEN TEA, 1 CUP .... caffeine: 40–60 mg

YERBA MATÉ, 1 CUP .... caffeine: 35–50 mg

GUAYUSA, 1 CUP .... caffeine: 30–66 mg

I also have a gene for mold susceptibility, and, sadly, coffee is one of the biggest sources of mycotoxins in our food supply. Studies show that 52 to 92 percent of green coffee beans are moldy.[6] Another option for the hard-core coffee drinkers is to swap moldy coffee for a low-mold version, such as Bulletproof (see Resources).

Similarly, alcohol worsens my memory and sleep, in part because I'm fifty and in part because I have that faulty methylation gene, MTHFR. See how problematic drinking is for me on many fronts? I love a glass of organic red wine, but more than two glasses per week keeps me awake, makes me retain water weight and feel bloated, and raises my liver enzymes so that I feel sluggish because I can't process my hormones and other internal chemicals well. I'm better off just skipping the sauce. The liver's job is to clear the gunk out of your body, and it can't do its job if you're keeping it busy with alcohol. That's why hangovers hit harder with age.

The trick is to discern which beverages serve you best, and when it comes to wine, be rigorously honest about it. It's common to deny health problems that arise from consuming alcohol, such as poor sleep, hang-overs, headaches, low energy, night sweats, hot flashes, and weight gain. Track your response with the objectivity of a scientific detective.

Goal: Settle your nervous system.

> **Protocol:** Reduce caffeine by half by switching from coffee to green tea or half caffeine/half decaffeinated; see if sleep improves.
>
> **Scientific Rationale:** Work around slow metabolism of caffeine (CYP1A2).

Goal: Avoid mold.

> **Protocol:** Stop drinking conventional coffee and choose low-mold brands.
>
> **Scientific Rationale:** Reduce exposure to mycotoxins and turn off the mold-susceptibility genes (HLA DRB1, 3, 4, 5, DQB1).

Goal: Improve microbiome; stop sugar-filled drinks, including juices that lack fiber.

> **Protocol:** Avoid sugar and artificial sweeteners.
>
> **Scientific Rationale:** Sugar and artificial sweeteners harm the microbiome and metabolism and may cause mitochondrial dysfunction.[7]

Goal: Lose weight and slow down aging.

> **Protocol:** Take resveratrol or drink moderate amount of organic wine (1 glass for women, two times per week).
>
> **Scientific Rationale:** Turn on longevity gene called SIRT1. Three glasses of alcohol or more is associated with a 15 percent higher risk of breast cancer.[8]

The Younger Protocol Elixir: Collagen Latte

I still drink caffeine, but now it's only the low-mold version and in conjunction with ingredients that help me metabolize it, giving me the benefits of caffeine without the drawbacks. Several years ago, when I launched my first book, *The Hormone Cure*, I started drinking collagen lattes. I was awake early for interviews on morning radio shows, and I couldn't quite bring myself to eat breakfast at five thirty, but a collagen latte was a creamy alternative. I didn't want to drink an old-school latte because dairy pokes holes in my gut and accelerates aging.

Once a collagen latte became my morning ritual, I noticed that I wasn't hungry until noon and my skin started to glow. So I looked into the data and was surprised to find that collagen, derived from gelatin, reaches measurable levels in the blood after oral consumption.[9] Collagen has the following benefits:

- Rich in antioxidants[10]

- Lowers blood pressure[11]

- Improves bone density[12]

Over time I've been experimenting with my recipe, which has involved regular coffee (such as Dave Asprey's low-mycotoxin Bulletproof beans or David Wolfe's low-acidity Longevity beans), decaffeinated coffee, and sometimes herbal tea made with chicory or dandelion. I may add a few drops of chocolate or English toffee stevia extract.

If you're unfamiliar with collagen, it's an easily digested form of protein that improves skin, hair, and nails. As you age, you break down more collagen than you make, leading to saggy skin, cracking fingernails, dull hair, and wrinkles.

How to Make a Collagen Latte

1 cup of low-toxin coffee, decaffeinated coffee, or tea

1 to 2 tablespoons collagen powder (See Resources for favorite brands)

Optional: 1 tablespoon of coconut oil or medium-chain triglyceride oil

Optional: 4 to 6 drops of stevia

1. Place brewed coffee and remaining ingredients into a blender.

2. Blend for 5 to 15 seconds until frothy like a latte.

Drink Organic Wine

If you do consume alcohol, red wine reduces mortality by more than 30 percent according to a meta-analysis of sixteen studies.[13] In fact, I suggest eliminating or severely limiting all alcohol but red wine.

I love wine, but a recent conversation forever changed the wines I drink. After I gave a speech at a locavore food event in San Francisco, I spoke with a winemaker who had a booth there. He told me about the agrochemicals used in the production of most commercial wines, a long list of additives that included sugar, tannin (beyond the tannins that occur in oak-barrel aging), acids, enzymes, copper sulfate, coloring agents, dimethyl dicarbonate (DMDC, a microbial-control agent designed to kill microorganisms), and fining agents. He pulled an amber bottle of copper sulfate pentahydrate, all corroded and toxic-looking, off a shelf in his booth and explained that it was routinely added to wine as a fining agent, designed to remove unpleasant odor related to sulfur, specifically hydrogen sulfide. Copper sulfate pentahydrate damages the kidneys of fish and is toxic to the liver and kidneys of rodents; in humans, it can cause liver and renal failure.[14] With my love of wine shaken, I continued the research on my own and found that grapes ranked number five on the Dirty Dozen list from the Environmental Working Group because of their high levels of pesticide residue.[15]

I learned there was another way. Wine is made from grapes, so it makes sense to choose wine made from organic grapes—as long as the wine tastes good. I tried organic wines and even biodynamic wines. Not all organic wines are biodynamic; biodynamic wines are made according to the principles of biodynamic farming, a form of organic agriculture with a tiny carbon footprint.

Rudolf Steiner (1861–1925) was the father of biodynamic farming and believed each farm should be a self-regenerative unit. Biodynamic practices oppose the use of fertilizers, pesticides, and herbicides. These farms prefer to use animal manure and compost rather than chemical additives. The focus is on the interdependence of the health of the soil, the growth of the plants, and the care of the livestock, so the crops reflect the *terroir* of each farm.

I haven't seen data to suggest that biodynamic wines are superior to organic wines, so both organic and biodynamic are probably good options. When you eschew additives, you get the full benefits of wine without the drawbacks. You may be wondering whether organic wine tastes good. It turns out that in addition to being better for your health and the planet, organic wine can also taste superb, although you may need to adjust your palate and experiment to find the best brands. A few of my favorite brands that taste similar to wines I used to love include Quivira, Preston, Truett-Hurst, Lambert Bridge, and Emiliana Coyam.

Drink one glass twice per week, which translates to about five to seven ounces per glass.[16] Don't forget that three or more servings of alcohol is linked to a modestly increased risk of breast and other cancers, on the order of 13 to 15 percent.[17] If you can't drink two glasses or less per week, skip it like I do. Which red wine is best? Pinot noir contains the highest concentration of resveratrol, a compound found in red grapes (and blueberries) that may offer several antiaging benefits, including prevention of type 2 diabetes, heart disease, and cancer[18] (data in humans are mixed).[19] Three studies even suggest resveratrol may mimic the longevity-enhancing effects (but unpopular tactic) of caloric restriction.[20] Effects are most impressive in people who are obese, with a BMI (a ratio of weight to height) greater than 30.

Resveratrol concentrations in pinot noir were higher than in other varietals across the world, except in the Trentino region in Italy, where the levels in cabernet sauvignon were higher. One possible explanation is that pinot noir grapes are harvested earlier than other types of grapes. Wines with the most resveratrol are from cooler regions. One study from Cornell found that pinot noir from New York contained more resveratrol than pinot noir from California.[21] So for wines highest in resveratrol, invest in

Pinot noir from California or New York

Cabernet sauvignon from California or New York

Italian Sangiovese

Australian Shiraz

French burgundy

Upgrade Yourself Orally

Shelley gave me a withering look. I was sitting in her dental-hygienist chair in Montclair Village, a neighborhood in Oakland. "Are you sure you're flossing daily and using the Sonicare? You have a ton of plaque, and I just cleaned your teeth three months ago." Her face brightened as an idea occurred to her. *Uh-oh.*

"Show me your flossing technique."

I obeyed, showing her how I position the floss at the right angle, taking care not to slice my gums between teeth. I slide up and down the adjacent teeth, just like she showed me last time (or so I thought).

"You get a C minus," she barked. I wasn't expecting that, and I started to wonder if she was perhaps a perfectionistic drill sergeant. "As you age, the saliva calcifies the plaque faster. Maybe's it's the minerals in your saliva or your microbiome in your mouth, but you need to upgrade the way you floss and raise the frequency as you age. Start flossing twice per day and using the Sonicare three times per day."

Really? I wasn't going to argue with her while she wielding sharp instruments around my mouth, but I decided that I'd prove her wrong with an exhaustive literature search.

Instead, I found support for all of her claims. And I suspect that no one is flossing right. Flossing fosters longevity, independent of brushing one's teeth, as does seeing the dentist at least twice per year (I go quarterly). If you don't floss, your risk of mortality is 30 percent higher, and if you see the dentist only once per year, you raise mortality by 30 to 50 percent.[22] If you wonder how that works, it may interest you to know the following:

- You have more than seven hundred species of bacteria in your mouth.[23]

- The tongue is a common place for biofilms to form—that's a group of microorganisms that behave like a gang, sticking to each other and adhering to the surface in a coating. They can cause bad breath, inflammation (gingivitis), plaque, cavities, and premature aging. Did you know that 22 to 50 percent of people have halitosis?[24] Me neither. Together, let's make the world a better-smelling place.

- More bad bugs in your mouth are correlated with thickening carotid arteries, a sign of atherosclerosis and precursor to stroke, which reduces blood flow to the brain.[25]

- Powered toothbrushes reduce plaque and gingivitis more than manual brushing in the short and long term.[26] Aim for a two-to three-minute date twice per day with your electric brush.

- In men, periodontal disease is more common and is linked to early atherosclerosis and heart disease.[27]

- Oil pulling is not as weird as it sounds. Swish 1 to 2 teaspoons of organic coconut or sesame oil in your mouth for five to twenty minutes five times per week. Oil pulling reduces gingivitis and the total number of anaerobic bacteria in your mouth (that's a good thing—makes your breath smell better while helping you live longer!).[28]

- Flossing can prevent periodontal disease after as little as one month of regular use.[29]

Floss at least twice a day. Buy a power toothbrush, and use it twice a day or more. Oil pull once a day.

Protocol for Week 1: Feed

Now that you know how food and drink can affect your genes, you are ready for your daily template for week 1. For the next seven days, follow these guidelines as closely as possible and be aware of the subtle yet profound change that is happening.

Sylvia from chapter 4 is my role model; every day she eats Mediterranean-ish with a lunchtime salad of lettuce, avocado, and a protein such as fish, chicken, a soft-boiled pastured egg, or beans. For dinner, she eats steamed vegetables with chicken or salmon and has nuts for dessert. She shops at her local market three times a week and always eats lettuce, tomatoes, avocados, cauliflower, carrots, sweet potatoes, yams, onions, garlic, apples, and whatever is in season, such as asparagus, green beans, or red peppers. Sylvia used to drink about three glasses of wine a week but gave it up a couple of years ago because it made her fall asleep too early. She flosses and brushes her teeth, with an electric toothbrush, three times a day.

Basic Rituals

- Eat green vegetables at least twice a day. I start my day with a green shake that contains one cup of green vegetables plus one scoop of a greens powder, which can be an easier way to make volumes of greens palatable. Check out the Recipe section in the appendix for some of my favorite recipes.

- Once or twice this week, consume either wild-caught fish, such as salmon, or plant-based sources of omega-3s, such as chia, flaxseeds, and purslane.

- Choose coconut, avocado, grapeseed, ghee, and olive oil. Use expeller-pressed organic coconut oil for cooking. Refined oils are usually processed with chemical distillation dependent on harsh solvents or they're hydrogenated, which creates trans-fats—avoid these fats. There are some good coconut oils available that are nonhydrogenated and refined using a natural, chemical-free cleaning process to make the oil better for cooking (higher smoke point), tasteless, and odorless, which may suit you if coconut is not your favorite flavor. Unrefined (virgin), organic, expeller-pressed coconut oil is your safest bet because no synthetic chemicals are used in its production.

- Don't use industrial oils, such as canola, corn, cottonseed, soybean, and sunflower. Steer clear of all hydrogenated or partially hydrogenated oils.

- Commit to eating at home this week, either cooking or assembling salads, soups, and meals. Start to develop an intimate relationship with food.

- Don't eat for at least three hours before bedtime. For me, that means no food after seven o'clock, or earlier if I'm intermittently fasting.

- Make a daily mug of hot tea, rich in polyphenols. Some of my favorites are matcha, Tulsi Sweet Rose tea, and Reishi tea, made from mushrooms, anise, licorice, and stevia. Or try Runa tea with guayusa, a native Amazonian tea leaf essential to making you *runa,* or fully alive. It has twice the antioxidants of green tea.

- Throughout the day, drink eight glasses of filtered water.

- Drink one glass of organic red wine two times a week, as long as you aren't triggered to drink more.

- Upgrade dental hygiene: floss twice and brush three times a day.

- Need tips? Go to TheYoungerBook.com for additional resources.

Supplement

Keep in mind that supplements are not meant to take the place of healthy whole foods in your diet, but they are an excellent complement to a diet that is missing certain micronutrients. Dose and effectiveness is crucial because supplements are an unregulated market.

- Take resveratrol. It has been shown to fight the effects of aging on a cellular level and to mimic the benefits of calorie restriction. The dose is 200 mg once per day.

Advanced Projects

- Start swishing with coconut or sesame oil, a practice called oil pulling. Take one to two teaspoons of coconut oil, place it in your mouth to melt (coconut oil is solid at room temperature, but melts at body temperature; sesame oil is liquid at both temperatures), and, keeping your lips closed, swish for five to twenty minutes. Don't swallow. Spit into your garbage or compost because it can clog your sink.

- Practice intermittent fasting by waiting twelve to eighteen hours between dinner and breakfast. For weight loss, perform intermittent fasting twice per week. For the aging benefit alone, perform once per week, starting this week.

- Make or buy bone broth, then drink a warm cup every day. Consuming bone broth is one of the best ways to replenish collagen in your body so that your hair regains luster, your nails, joints, and teeth strengthen, and your gut seals over the leaky junctions between cells (see Recipes).

- Measure and reset blood sugar. Sugar, stress, and bad genes can throw off your blood sugar, and the best way to know is to measure your fasting blood sugar with your doctor. See Resources for details.

Recap: Benefits of Week 1

Starting this week, you are applying a new kind of science to your genes and changing how they respond to environmental inputs such as nutrient-dense food, drinks, and targeted supplements designed to re-direct your genes to a longer healthspan. You've ratcheted up the level of antioxidant foods and reset the right dose of alcohol so that you don't overburden your methylation and oxidative-stress genes. For you, two glasses this week may be the proper dose, but be truthful about your ability to stop. You've also started prevention of cancer and other degenerative diseases. You've boosted total fiber, which is linked to successful aging. You'll continue the feed protocol for the full seven weeks.

Bottom Line

Eating and drinking food that improves your exposome will enhance the immediate and long-term math of aging in your body. Plus, when you eat well, your immune system fights off illness better, you elevate your brain function, you lose weight, and your energy soars. Upgrading your food is the best way to get a makeover, internally and externally. The visible and invisible benefits are unsurpassed by any other lifestyle/epigenetic change you can make. Week 2 will help you redouble your energy and brain function from another angle of epigenetics: sleep and whether you're getting the right amount.

CHAPTER 6

Sleep
WEEK 2

Think in the morning. Act in the noon.
Eat in the evening. Sleep in the night.

—William Blake, English poet and painter

I undersleep. I'm not alone: Americans today sleep about three hours less per day than they did prior to the Industrial Revolution a hundred and fifty years ago. Hectic lives, caffeine, grueling desk jobs, artificial light at night, and ubiquitous screen time are contributing to less sleep and more aging. When you disrupt your circadian clock, your cells get angry. Your circadian clock is a molecular timekeeping system that's present in nearly all your cells and strongly influenced by genetics and the environment, particularly sleep.

You may think that undersleeping is not a problem for you, and boast that you are a naturally short sleeper who performs just fine on six hours of sleep. The truth is that only 3 percent of the population has the short-sleep gene, known as hDEC2-P385R (or DEC2 for short), one of several genes in charge of circadian rhythm. People with this gene polymorphism need fewer than the recommended seven to eight and a half hours per night.[1]

Leo Tolstoy famously said that all happy families are alike, but each unhappy family is unhappy in its own unique way. The same is true for sleep. Happy sleep is remarkably similar from person to person; you awaken refreshed, ready to face your day with aplomb. Unhappy sleep is awful in unique ways. It wreaks havoc on your internal biochemistry. Not only are you at greater risk of having accidents, your short-term memory becomes toast along with your focus and attention. You're more likely to overeat and feel depressed. Sleep deprivation is linked to obesity, diabetes, heart disease, stroke, and early mortality. Makes me want to go to bed right now!

One sleep expert, Charles Czeisler of Harvard Medical School, says that people who sleep fewer than five hours a night for five consecutive years have a 300 percent greater risk of hardened arteries. While hardened arteries are hard to reverse, with just a few weeks of restorative sleep, you can lower your blood pressure and improve cellular repair, which may indirectly help.[2]

I know these dire sleep statistics are true from personal experience. In 2012 when I was working on my first book, *The Hormone Cure,* I sat a lot to write and racked up a gigantic sleep debt. At the time, I was seeing patients full-time in my functional medicine practice. Then I would come home to my second shift: I'd make dinner, supervise homework, get emotionally current with my family as fast as possible. Finally, third shift: late nights and early mornings of writing after the kids went to bed and before they woke up. Soon my good behaviors fell off; I skipped yoga classes and weekend runs with friends. I drank more coffee. I needed wine to wind down at night so I could sleep. I all but abandoned my poor husband.

I got the book done on time but at great personal cost. During that time, my family felt neglected, and my husband called me a workaholic. (He was right.) I slept four to six hours per night. I gained weight, but, even worse, I became increasingly stressed and stiff and felt older than my years.

My wakeup call came when I tested my telomeres, those markers of biological aging. The results shocked me. At age forty-five, I had the telomeres of a sixty-five-year-old woman. I realized no amount of pills or injections could save me; I needed to repair my body and my telomeres the old-fashioned way, with sleep and movement, which you'll read more about in this chapter and the next.

Sleep is like a gifted housecleaner who comes into your home while you're away and tidies everything up, bringing peace of mind, clean surfaces, and folded bedsheets. In short, a fresh start of biological reorganization. Beyond tidying up, the repair that occurs during sleep works wonders in every realm of your life. It's as close to a panacea as you'll find. Master your sleep and you'll find it easier to balance your hormones, muster the willpower to make good choices, and, ultimately, lengthen your healthspan.

With specific techniques outlined in this week's protocol, you'll regulate the circadian rhythm genes, such as Clock, the longevity genes SIRT1 and mTOR, and you'll turn off the bad genes associated with a dysfunctional sleep-wake cycle.

Why It Matters

Even though you spend about one-third of your life sleeping, most people don't think much about it or get enough of it. The National Sleep Foundation suggests adults get seven to nine hours of sleep per day, but nearly 30 percent of people in the United States get six hours or less. Sleep is an important window into your world, from biology to health. If you're one of those people who scrape by with fewer than seven hours of sleep a night during the week, hoping to catch up on the weekend, I have good news and bad news.

Good news first: you can change the way you sleep, starting tonight. We sleep so much less these days thanks to the electric lightbulb, TV screens, computer screens, and other artificial lighting, all of which distract us from the delicate signals of the body's inner clock.

Bad news: you can't catch up on the weekend. If you sleep four hours per night for five days, you have a twenty-hour sleep debt, and you can't make that up in a weekend. Other strategies can be helpful for this problem, which is known as social jet lag, or the discrepancy between the midpoint of your sleep on nights when you don't have to wake up early the next day versus the midpoint of your sleep on nights when you do have to get up early. One tactic is napping. A twenty-minute nap can be as healing to the

Sleep by the Numbers

- Sleep deprivation makes your genes go rogue; 97 percent of rhythmic genes become arrhythmic, which is a dangerous alteration to your DNA because it may lead to diseases such as cancer.[3]

- Going to bed four hours later than normal, when repeated for three consecutive days, leads to a sixfold reduction in circadian genetic messages.[4]

- Sleep deprivation changes the expression of one in three genes.[5]

- Sleeping five hours or fewer a night equates to aging an extra four to five years, according to the Harvard School of Public Health.[6]

- More sleep isn't always better. One study showed that women fare best on cognitive tests at seven hours of sleep. But increasing sleep beyond eight hours is associated with *lower cognitive scores*, equivalent to a five- to eight-year increase in age.[7]

body as an hour of nighttime sleep, although results may vary.[8] According to research on long-lived populations, napping lowers stress levels.[9] Yes, I just gave you permission to schedule adult naptime for yourself.

That's not the only downside to sleep deprivation; when it comes to eating, people who don't sleep enough have an imbalance of two important satiety hormones: ghrelin and leptin. If you sleep five hours a night or less, you have more ghrelin, which tells you to eat, and less leptin, which tells you to stop. In short, you're hungrier when you sleep less.[10] The problem with ghrelin and leptin, combined with disruption of other key healthspan hormones such as melatonin, cortisol, insulin, and growth hormone, may be why sleeping less is associated with getting fat.[11] Not only that, but your willpower, which you can think of like a fuel tank, doesn't get refilled. Of course, sleep debt is related to many other unfortunate outcomes; your circadian rhythm regulates 15 percent of your genome, so getting your sleep-wake cycles lined up for greatness is essential to your health. If you blow it off, you may lack the energy to exercise and run the risk of promoting your chance of certain types of dementia, including Alzheimer's disease.[12]

By now you have a pretty good idea of my mediocre genetics, so no surprise that I have a polymorphism of the Clock gene that requires me to sleep eight hours per night in order to lose weight. If I don't, I have high blood levels of ghrelin, which tell me I'm famished. So I'm very motivated to turn on my Clock gene and get it working for me, not against me. You probably want to do the same. (There are more work-arounds besides sufficient sleep, such as consuming protein—a protein shake or collagen latte—in the morning soon after waking up, which helps the Clock gene to keep ticking properly.)

Nate, the Eighty-Year-Old Nordic Ski Racer and Triathlete

I was walking on my treadmill desk, editing this book, when a photo arrived via text from a friend. The photo showed her father, Nate, midstride in a Nordic skiing race in Tahoe with the number 315 on his chest. I wasn't surprised.

Several summers ago, I visited my friend in Tahoe and met Nate. One morning, as I puttered around their kitchen, pouring coffee, Nate came in and told us he had just completed the Lake Tahoe triathlon. When he finished, he felt so good that he ran through the entire course two more times—while I slept. I stared at him. I had never heard of such a thing. He locked eyes with me and said, "Always incorporate racing into the forecast. The cart will follow the beast." I was in the presence of a super-ager.

Nate has completed thirteen Ironmans and fifty marathons, usually winning his age category (unless a seventy-year-old beast passes him; he hates that). I got curious and asked about his daily habits. He explained that he started his day with coffee. Drank eight glasses of water a day (apparently beer counts as water). Played chess every night. Brushed his teeth three times per day. Took at least one fifteen-minute nap every day. Went to bed by ten. Slept at least seven hours. Now Nate is eighty and looks younger than the sixty-year-olds he beats on the racecourse. Resting heart rate? Fifty.

Official Definition of Sleep Deprivation

With all this talk about sleep deprivation, you may wonder how it's defined. Can one person thrive on seven hours while another needs nine? Yes. In short, sleep deprivation is when a person gets insufficient sleep to maintain wakefulness during the day, and it varies from one individual to the next.

The way that sleep researchers know a person is sleep deprived is via a psychomotor vigilance task. (See Resources.) During the test, a bull's-eye appears on a screen, and the subject has to press a button. When someone is sleep deprived, after a few minutes, he or she loses the ability to focus on the task and forgets to press the button. Some people can sleep for seven hours and pass the test, and others need eight hours of sleep or their performance on the test declines.

An Insomniac Overcomes Her Sleep Problems

When I got Rosalie's test results back, my jaw dropped. I conjured up my mental image of Rosalie: piercing blue eyes that twinkle with her dry wit, short no-fuss haircut, supple body and joints. Rosalie is stunningly young for her age. She's seventy, but her telomeres say she's fifty, my current age. I scratched my head and considered why she was doing so well at the art of aging.

You may think of the usual reasons, such as an attitude of gratitude or a lack of perceived stress.

No, actually Rosalie had a demanding career; she was a journalist who'd risen rapidly through the ranks to become a leader in her field. She experienced the near-constant stress of deadlines, zero accommodations for being a working mom in a traditionally male field, cross-country moves, a divorce—the usual high stressors.

Did she get fantastic sleep, seven to eight and a half hours every single night?

Well, yes and no. Here's where things get interesting. Rosalie has struggled with sleep for twenty years, starting in her midfifties with menopause. She began acupuncture, which helped a little. Then she tried the Integrative Medicine Center at Griffin Hospital in Derby, Connecticut, which was run by the famed David Katz, M.D. Again, it helped a little. When Rosalie moved to the Bay Area, she came to see me and we balanced her hormones: cortisol, estrogen, progesterone, and thyroid. We addressed her stress system. Her sleep got better. Since then, she's added cannabidiol oil under her tongue before bedtime, and she also listens to a CD of Benedictine monks chanting. All of these changes have led to incremental improvements, but remarkably, her telomeres have managed well without the perfect sleep number of seven to eight and a half hours a night.

I mention Rosalie not because I want you to start cutting corners on sleep, but because she has taken the baby steps that add up to a longer healthspan. What I've also noticed is that she is endlessly curious, which

Sleepers Barely Work

Sleeping pills, also known as sleepers to physicians and pharmacists, aren't the answer. While 5 percent of Americans take sleeping pills, a number that's doubled in the past twenty years, they add only twenty to thirty-seven minutes of sleep time, and that sleep is not always the best quality. Sleeping pills allow you to fall asleep faster but at a high cost—they're addictive; they cause memory loss, daytime sleepiness, and brain fog; they worsen sleep quality; they're associated with cancer; and they raise mortality, even if you take as few as twenty pills per year.[13] The top FDA-approved non-benzodiazepine sleeping pills may owe up to 50 percent of their benefits to the placebo effect.[14] Researchers are not sure if sleeping pills cause cancer or raise mortality or if these problems are associated with the lack of sleep and not the pills, but I can confidently speculate that lack of sleep creates a "bad neighborhood" in your body—even one night of fewer than seven hours of shut-eye is linked to high serum levels of cortisol, which sets off unnecessary inflammation in the body.

may end up being as important as flossing or keeping in touch with your close friends in predicting longevity. She has a core sense that there is a lot of work to be done, whether that's playing a key role in raising her grandchildren or jumping in as an activist in her community. Perhaps most important, she has a settled philosophy about the fundamentals in life. She doesn't expect a miracle drug that will help her with sleep. Yet, when you're around her, you relax a little, because she has figured out her role in the world. It's refreshing. Not just for me, but for her telomeres.

Your Mission, Should You Choose to Accept It

Just as we want to switch on the good sleep genes so your circadian rhythm is happy, we want to avoid the poor sleep—even a single sleepless night—that can flip the switch of your genes and put you at risk of disease.[15] One sleepless night is linked to poor driving performance and reaction times. When you miss too much sleep, you may end up in one of the ten thousand to twenty thousand car accidents caused by a driver falling asleep behind the wheel.[16] Don't get me started about the poor work performance and decreased ability to learn. Beginning this week, the key is to comprehend how much sleep you require to feel your best and slow down aging. Here are a few questions that we will consider so we can optimize your sleep, energy, and aging process:

- How many hours of sleep allow you to wake up feeling refreshed and alert?

- Does being exposed to more daylight, especially before noon, help you sleep better at night? (This practice raises your internal melatonin levels at night.)

- If you skip caffeine and alcohol, does your sleep improve?

- Does avoiding that bad boyfriend ALAN (artificial light at night) improve your sleep?

- Does sleep improve with quiet activity, lit by candles, the hour before sleep? For example, a soothing yoga practice or guided visualization? How about listening to chanting monks, like Rosalie?

Science of Week 2: Sleep

Sleep problems are often unrecognized, underestimated, and, frankly, ignored by mainstream medicine and employers. That means we need to help ourselves by becoming sleep evangelists and paying attention to what throws off our sleep-wake cycles. Your circadian rhythm and sleep can get out of whack any number of ways. Here are a few proven circadian disrupters, some of which are bidirectional in terms of cause and effect:

- Shift work—such as when you work nights—is now considered by the World Health Organization to be a probable carcinogen.[17]

- Chronic stress. Under chronic stress, some people maintain normal function while others become vulnerable to mental illness, ranging from addiction to depression.[18]

- Caffeine[19]

- Jet lag and time-zone changes[20]

- Psychiatric disorders[21]

- Artificial light at night (ALAN) from lamps and electronics[22]

- Space travel, for all of you aspiring astronauts and followers of Elon Musk and Richard Branson[23]

So endeavor to bring your sleep debt to a zero balance.

Architecture of Your Sleep

While quantity is important, quality matters too. There are two types of sleep: rapid-eye-movement (REM) and non-rapid-eye-movement (NREM). NREM sleep is further divided into stages 1, 2, 3, and 4, indicating a spectrum of relative depth.[24] During this week—and for the rest of your life—you want to improve and lengthen slow-wave sleep (deep sleep, stage 3) because that's when the heart rate lowers; organs, muscles, and bones rebuild; energy and willpower reserves get replenished; and your immune system resets. You need your REM sleep for emotional regeneration and NREM slow-wave sleep for physical regeneration. All phases of sleep are important. You will benefit the most if you make your pattern consistent night after night in regard to onset, depth, and duration.

Your sleep is most restorative when it consists of all stages occurring in ninety- to hundred-and-twenty-minute cycles during the night. During the stages of sleep, you reap the following benefits:

- Growth hormones and melatonin levels rise while cortisol decreases. (Sleep deprivation is associated with higher cortisol levels in the afternoon and evening and with elevated blood sugar, similar to what is observed in the aging process.[25])

- Memory is stabilized and expanded, both at night and in naps.

- Slow-wave sleep enhances declarative memory, the type of memory that is consciously recalled, such as verbal or factual knowledge that you can state in words.

- REM sleep is most important for emotional recalibration and formation of nondeclarative or procedural memory, which is skill-based (for instance, learning how to ride a bicycle).

In sum, sleep is essential to slow down aging because that's when growth hormone does its repair work in the body. Skipping one phase or more ages you faster and raises your risk of early death. All hormones are released in a circadian rhythm, and sleep-wake cycles set your rhythm so that you can make melatonin and growth hormone at night. Melatonin controls more than five hundred genes in the body, including the genes involved in the immune system, so managing your melatonin is a sound investment. When you cut sleep, you reduce your levels of growth hormone, and you may be less able to repair injuries and more likely to accumulate belly fat.

Health Problems Associated with Sleep Deprivation

Immune Function

If you sleep fewer than seven hours, you compromise your immune system. In one study, subjects were limited to seven hours of sleep or fewer for one week and then exposed to the common cold virus. Short sleepers were twice as likely to catch a cold.[26] That means you may be able to sleep away your risk of catching a cold starting this week.

Cancer, Diabetes, Stroke, and Heart Disease

Since poor sleep is associated with disrupted melatonin production, it may put you at greater risk of cancer.[27] We see this most in women who work night shifts,[28] such as nurses and flight attendants, but debate continues about the strength of the association.[29] People who work at night may be at greater risk for cognitive decline,[30] cardiovascular disease,[31] stroke,[32] diabetes (especially for women),[33] and certain types of cancer such as breast, ovarian, and prostate.[34] Most of these risks take a while to accumulate, typically fifteen to twenty years. In 2007, the International Agency for Research on Cancer (IARC) stated, "Shift work that involves circadian disruption is probably carcinogenic to humans."[35] In one study, short sleep duration and frequent snoring were associated with significantly poorer breast cancer survival. For those women who slept six or fewer hours per night, the risk of death doubled.[36] Obviously, I slept less during many years of my medical training to become an ob-gyn, so now preventing sleep debt is a top priority for me.

Cognitive Decline

Sleeping too little *or* too much is tied to worsening brainpower. In fact, less sleep makes your brain function slip, and cognitive decline worsens your sleep.[37] For every thirty minutes extra it takes women to fall asleep at night, the chance of cognitive impairment goes up by 13 percent.[38] *What?* Plus women who nap two or more hours a day raise their risk of impairment, so keep those naps short (around twenty to sixty minutes).

Not far from where I was born in Maryland, there's an ongoing investigation called the Baltimore Longitudinal Study of Aging. Started in 1958 at Johns Hopkins University, it's the longest scientific study of aging. Recently, the group found that fewer than five hours of sleep (or poor sleep) is tied to higher levels of beta-amyloid, that nasty abnormal protein that can build up in your organs. It gums up the structure and function of your tissues and collects in the brain, eventually forming plaques and leading to memory and cognitive impairment. Not surprisingly, beta-amyloid accumulation can disrupt your NREM sleep and memory.[39] So disturbances with your sleep-wake cycle may precede and even contribute to Alzheimer's disease.[40]

Early Birds and Night Owls

You're probably wondering what all of this science means if you love getting up early or staying up late. You have a genetically determined preference to be an early bird or a night owl, which represent different circadian phases.

When I fell in love with my husband, I didn't notice the time of day when I was with him. But after fifteen years with him, I know that I get up earlier and go to bed earlier. Turns out that a woman is more likely to be a morning person, also known in scientific circles as a morningness chronotype, whereas men are more likely to have the eveningness chronotype.[41] A *chronotype* is a biologically encoded difference toward morningness or eveningness, with a continuum in between of intermediateness (okay, I made up that last term).

Everyone has a circadian rhythm, but most people beat at a slightly different rate with an average length of twenty-four hours. Early birds have a circadian rhythm that's shorter than twenty-four hours; night owls' are slightly longer. Generally, women beat six minutes faster than men.[42] That may seem minor, but imagine a clock that gains six minutes a day. After ten days, women may be off by an hour compared with men, so a periodic reset is important (unless you're living alone in a cave). Even if you live with someone of the same gender, you may have different chronotypes.

Daylight is the primary cue, or zeitgeber, that aligns your inner twenty-four-hour day/night clock with the environment. That means you may need to modify your exposure to daytime and nighttime light in order to live harmoniously within your own circadian rhythm and with someone who is your opposite, as I'll describe later in the advanced projects for week 2.

Want to know your chronotype? Take a weeklong vacation—go to sleep when you want, and awaken when you want, ideally free of the influence of caffeine or alcohol. By the end of the week, you'll know your chronotype. Or, far less fun, you can answer the question on the Munich ChronoType Questionnaire.[43]

Sleep and Vitamin D: A Perfect Match

For those of you looking to improve your sleep, here's an easy win. Vitamin D is good for more than just bones. According to experts, vitamin D appears to have direct brain effects on your regulation of sleep, specifically in the diencephalon (the part of your brain that contains the hypothalamus and regulates hormones) and brain stem (trunk of the brain). Some hypothesize that sleep disorders have risen to epidemic levels because of widespread vitamin D deficiency, and I agree.[44] You may be at even greater risk with a faulty vitamin D receptor gene like mine (see appendix). Keep your vitamin D levels in your blood between 60 to 90 ng/mL. If you're still skeptical, allow me to share more data to convince you.

- Lack of sleep, disrupted circadian rhythms, and low vitamin D levels can impair healing and repair.[45]

- Low serum vitamin D may make you take longer to fall asleep.[46]

- In a group of 459 postmenopausal women, higher vitamin D concentrations were associated with better sleep maintenance, meaning that it was easier for people to stay asleep.[47]

- Older men (age sixty-eight and up) with low levels of vitamin D in the blood were likely to have short sleep duration, poor sleep

Sleep Shampoos Your Brain

There's a wash that runs through your brain at night, while you're sleeping, that removes damaging and toxic molecules associated with neurodegeneration, basically the decay of your brain.

Here is how the *brain shampoo* works. During sleep, the space between brain cells expands 60 percent more than when you're awake. This allows the brain to flush out built-up toxins with cerebral spinal fluid (CSF), the clear liquid surrounding the brain and spinal cord. It's called the glymphatic system. One study showed that while you sleep, your glymphatic system might

efficiency, and increased interruptions in their sleep.[48] In fact, subjects with levels of less than 20 ng/mL had double the risk of sleeping fewer than five hours than men whose levels were greater than 40 ng/mL.

- Low vitamin D levels are associated with sleepiness.[49]

- Vitamin D has hormonal, neurological, and immunological influences on pain in the body, playing a key role in chronic pain and associated problems such as insomnia.[50]

- Low vitamin D is associated with restless leg syndrome, which may impair your sleep.[51]

Other nutrient deficiencies, including low levels of vitamin B_9 and B_{12}, are also associated with sleep problems.[52] For B_9 (folate), eat black-eyed peas, lentils, spinach, cauliflower, peas, and okra. For B_{12}, eat grass-fed beef liver, mackerel, sardines, and salmon. Keep taking your B vitamins for reasons described in the introduction and chapter 4 and continue them throughout the protocol for better sleep.

clear beta-amyloid, that protein associated with Alzheimer's disease, more rapidly than it does while you're awake.[53] Your glymphatic system works best when you're sleeping on your side, not on your back or tummy.[54] I use a pillow between my bent legs to pin myself in a side-lying position because I get a better brain shampoo and it helps decompress my low back.

The Dark Side of Blue Light

Doesn't all this sound good enough to make you want to cultivate your sleep? The problem, though, is that modern life—more specifically, artificial light—often interrupts us. Anton Chekhov, the physician and playwright, is famous for this line "Medicine is my lawful wife and literature my mistress; when I get tired of one, I spend the night with the other." Similarly, I love to curl up at night, around nine, after the kids are in bed (and hopefully not looking at their smartphones!) with a good book on my

From Dr. Sara's Case Files: Vitamin D and Me

Vitamin D is important for the efficient trafficking of calcium in your body, which helps keep your bones strong. I inherited a gene for a faulty vitamin D receptor, which means my body doesn't absorb and transport vitamin D well, so I tend to have low levels of vitamin D in my blood.[55] This puts me at greater risk for accelerated bone loss, osteopenia, osteoporosis, multiple sclerosis, and certain malignancies such as colorectal cancer.

When I learned about my insufficient vitamin D genes, in 2006, I began taking larger doses of vitamin D. The Institute of Medicine recommends 600 IUs of vitamin D per day,[56] but that low of a dose might put me at a greater risk of bone loss and keep my vitamin D receptor gene turned off. Taking a higher dose of vitamin D (I take 5,000 IU per day) helps me prevent excess bone loss and osteoporosis and maybe even the dowager's hump that you commonly find in elderly women. A higher dose of vitamin D is an *epigenetic change* that allows me to keep the switch *on* for my body's use of vitamin D.

iPad. Is that a good idea? Maybe not. My habit might be robbing me of getting all five stages of sleep and contributing to my risk of cancer, diabetes, heart disease, and getting fat.

The fastest way to throw off your biological clock is to blast your eyeballs with ALAN. The blue wavelengths of digital screens are fine during the day because they improve your attention, reaction times, and mood, but they turn on you after sundown. Not all colors are the same; the blue light emitted from digital screens, such as smartphones or e-readers, suppresses melatonin production more than other colors of light. Teens are particularly vulnerable.[57] The brighter the light, the bigger the impact on melatonin production, but even small amounts of screen time can make you sleep deficient.

Screens aren't the only ALAN culprit. Last year, my husband replaced every single lightbulb in our home with a more ecological fluorescent or LED light. We're talking more than a hundred curlicue compact fluorescent bulbs. Alas, what's good for the planet may not be good for our health; they are more energy efficient but emit more blue light than the old-school incandescent lightbulbs. Oy!

So if you live with a tree hugger, like I do, or you're a night owl, a shift worker, or an astronaut, invest in indoor glasses that block blue light, and make sure you shut off those screens at least an hour before bed. (See Resources.)

Protocol for Week 2: Sleep

Here is your template for week 2. For the next seven days (and for the rest of the full protocol), follow these guidelines as closely as possible. You should start noticing that you feel emotionally and physically refreshed and handle stress better. Within a few weeks, your immune system will be stronger, and you'll contract common illnesses less and less.

Start by taking a psychomotor vigilance test to assess your sleep debt. It's a simple test that you can perform online or on your smartphone. (See Resources for details.)

Record score here: _____ and date: _____.

Basic Rituals

- Keep your bedroom free of all electronics; if that's not possible, keep them at least five feet away from your body an hour before bed. This action will improve the quality and quantity of your sleep.

- For women forty and older:

  - Maintain a nighttime room temperature of sixty-four degrees or cooler to minimize temperature disturbance. This

Rationale for Tightening Up Your Sleep

Now you know why it's so important to get the best sleep possible, but I must caution you about using your weekends to bask in hours of sleeping in. Getting too much sleep can cause its own set of problems. All large epidemiological studies offer the same bad news: sleep too much or too little, and you will die early.[58] One of the largest studies of sleep duration and mortality followed 1.1 million people, including 636,095 women age thirty to a hundred and two, for four years. Mortality has a U-shaped association with sleep duration, meaning that moderate amounts are optimal but low or high levels can be harmful (the same is true of exercise, food, and sunlight exposure). It showed that women with the highest mortality slept fewer than seven hours a night or more than eight and a half. These women were also older, heavier, and more likely to be ill. On top of that, altered sleep is associated with more alcohol use and less exercise. I'm better off cutting out alcohol and exercising more in terms of sleep and slow aging. *Sigh*.

Long sleep, like short sleep, is associated with depression in postmenopausal women, which makes me curious about the long sleepers. It's a which-came-first-the-chicken-or-the-egg situation. That is, are they sleeping longer because they're depressed, or are they depressed because they're not getting out of bed in the morning?[59] In sum, there may be a benefit to a slight reduction in sleep, as long as sleep quality is good. When your stress is high and you feel exhausted, aim to stick within the sweet spot of seven to eight and a half hours of sleep.

can be challenging and expensive during summer—just do your best.

- – Address hot flashes and night sweats. You may need to consider short-term bioidentical hormone therapy to improve sleep. Specifically, natural progesterone, 100 to 200 mg, improves sleep in women in perimenopause and menopause.

- Avoid stimulants. No coffee and no anxious people. Both overstimulate your nervous system, making it harder to sleep. Yes, I'm suggesting that you reduce your exposure to people with anxiety, especially if you're a highly sensitive or empathetic person. You are the average of the five people you hang out with. Make them Zen-like so that you don't perseverate on how to please them when you go to bed.

- Exercise in the morning, or at least before one in the afternoon. If you must exercise later, notice if your sleep duration and quality declines and adjust accordingly.

- Create a sleep-conducive environment that is dark, quiet, comfortable, and cool. Your room should be so dark that you cannot see your hand when holding it in front of you.

- If you slept fewer than seven hours the night before or if you feel tired, take a nap for at least twenty minutes. At least once this week, nap for twenty to thirty minutes.

- Notice if you're sleeping more than eight and a half hours each night. Set an alarm if needed.

- Expose yourself to lots of bright light during the day, which will improve your ability to sleep at night as well as your mood and alertness during the day.

- Take a mini-savasana. With savasana, or corpse pose, you lie on the floor on your back with your palms facing up and your legs hip-width apart, and you let go, relaxing every muscle in your body. It helps me chill out, not just after a tough yoga class but also during the day. My body has an extreme response to the daily bustle of life, and taking a mini-savasana for two to five minutes allows me to shake off the day.

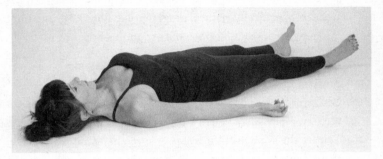

- Go to bed before ten, or at least thirty minutes before your usual bedtime, and sleep seven to eight and a half hours. Go to sleep at the same time every night, seven nights a week. Get up at the same time every day, even on the weekends. Binge sleeping doesn't work because you lose the normal sleep architecture and sleep quality suffers. Try to keep sleep architecture almost exactly the same every single night this week. Aim to be in bed before ten if possible (night owls, do your best to go to bed earlier than usual), ready for sleep, not chatting with your partner or answering last-minute e-mails.

- Reset your sleep-wake cycle.

- Eat your carbs at dinner. Not cupcakes but quinoa, sweet potatoes, and yucca, which are digested slowly and don't raise blood sugar excessively. Don't eat for three hours before bedtime, so your curfew for food is 7:00 P.M. Eating your carbs at dinner will help activate the genes of weight loss, including the ones that code for leptin, ghrelin, and adiponectin.[60]

- Limit your use of screens and ALAN for three hours before bedtime, which will be 7:00 in most cases. Turn off electronics, such as TVs, computers, phones, and tablets, at least one hour before bedtime. That means no more television in the bedroom or falling asleep to TV.

- Design a regular and relaxing bedtime ritual, such as soaking in a hot bath with Epsom salts or listening to soothing music; begin an hour or more before you expect to fall asleep.

Supplements

Vitamin D. Correcting your vitamin D deficiency will do more than strengthen your bones—it may help you sleep better. Vitamin D affects at least three thousand genes. Generally, I recommend about 2,000 to 5,000 IUs/day, but the best strategy given the multiple genes involved in vitamin D metabolism is to track your blood level over time. Take enough vitamin D to keep your level 60 to 90 ng/mL. The target range shifts slightly based on research updates, so the range may be adjusted in the future. If your serum vitamin D is usually less than 60 ng/mL, don't worry, as it's unclear whether higher levels will benefit your telomeres (higher levels have not been studied for correlation with telomere length).[61] As you'll find with many nutrients, there is a U-shaped curve between vitamin D and health, so too little is bad, and too much is bad too. You want just the right amount for your biology.

Advanced Projects

- Use bright-light therapy. Sit close to a special light box for a certain amount of time each day. The light box emits visible light that mimics outdoor light, important for regulating your circadian rhythm.[62] It may help you sleep later in the morning or fall asleep earlier at night, depending on how you use it.

- Wear amber-colored glasses after the sun goes down. They're the new reading glasses for women who want to age in reverse, but get the indoor variety that blocks blue light (but not other colors). Wear them from sundown to bedtime so that you make more melatonin and sleep more soundly.[63]

- Assess your deep sleep. There's a gene for deep sleep, and the variant helps you sleep more deeply. The SNP ID is rs73598374. My husband has the typical gene (no variant, C;C) whereas I have the deep sleep variant (C;T). If you compare data from our sleep trackers, you'll find that my deep sleep is twice his on the same night. To assess your deep sleep, try one of the trackers listed in Resources.

- Get your melatonin tested, or try this inexpensive melatonin reset. My melatonin level is low-normal, so periodically, I will perform a melatonin reset by taking a tiny dose of melatonin (0.4 mg) four hours prior to bedtime. After I take this small dose, my melatonin level initially spikes and then starts to decline. My body detects the decline and causes my pineal gland to make more melatonin so that, four hours later, I can sleep more soundly.

- What to do with circadian opposites. Since women's circadian clocks beat six minutes faster than men's, and women are more likely than men to be early birds, you may need to adjust your life to your circadian opposite.[64] You can shift your built-in predisposition by putting cues in your environment; usually, getting up an hour earlier or later than usual to accommodate your partner isn't difficult. Bright light in the morning can cue a night owl to wake up, and blackout shades at night can cue the night owl to wind down. Pumping iron in the bright lights of a gym later in the day can also cue an early bird to go to bed later.

Your Daily Routine

The next page shows a daily routine that includes the basic rituals from weeks 1 and 2. This is a typical day for me—a normal day with better choices. Modify as you see fit.

A TYPICAL DAY IN THE YOUNGER PROTOCOL: DR. SARA

6:00 A.M.	Wake up, floss, and brush teeth with electric toothbrush
	Take supplements on empty stomach
6:05	Brew and drink green tea collagen latte
6:10	Meditate for 10–30 minutes
6:45	Breakfast shake or skip if intermittently fasting
7:30	Take kids to school
8:00	Exercise (barre interval class, yoga, or brisk walk)
	Sip/drink branch-chain amino acids (see Supplements)
	Drink 1 liter filtered water afterward
10:00	Work
Noon	Lunch of leftovers (e.g., chicken, broccoli, finger salad of raw veggies) or green shake
	Brush and floss
	After lunch walk and work (about 2–7 miles on my treadmill desk)
	Drink 1 liter filtered water
3:00 P.M.	Pick up kids, drink more water
3:30	Drag kids to sauna at the gym, if they're game (you'll see why in chapter 8)
4:30	Meditate for 10 minutes or call a friend
5:00	Prepare dinner (make double recipe so there's leftovers for lunch)
6:00	Dinner
7:00	Me time or family time – when you metabolize food and the day
9:00	Shut down backlit screens and artificial light at night
	Brush and floss
10:00	Go to bed, lights out

Recap: Benefits of Week 2

Improved sleep gives you many benefits: you can reset your hormones, strengthen your immune system, and find it easier to lose weight. Less sleep ruins your blood-sugar control and raises levels of stress hormones, such as cortisol. It turns your hunger button to the on position perpetually, which is not good for improved lean body mass with age.

Optimal sleep will activate growth hormone so that you are less likely to gain belly fat and better able to repair your muscles at night. Immediately, you'll notice better memory and more resilience in fighting colds. Long term, you will find that your risk of diabetes, high blood pressure, and obesity (all of which are associated with mortality) decreases. You'll also notice less inflammation and stiffness, all because of resting more!

Bottom Line

The Dalai Lama claims that sleep is the best meditation, and I agree. While sleep may be largely ignored by mainstream medicine, that doesn't mean you need to ignore it too and pay the consequences later. A small effort now can yield a large epigenetic change. Molecular adaptations triggered by your environment can create improvements in your longevity and weight genes. So make bedtime as important as what you eat, and get the right dose of sleep now.

CHAPTER 7

Move

WEEK 3

From an evolutionary perspective, exercise tricks
the brain into trying to maintain itself for survival
despite the hormonal cues that it is aging.

—John Ratey, *Spark: The Revolutionary New
Science of Exercise and the Brain*

I wake up to pee and stumble in the dark to the bathroom. I glance at my watch and see it's 5:30 A.M. I'm relieved that I can climb back into bed for another hour of sleep. For a moment, I think about my close friend Allison, who lives just a few miles away. No doubt Allison, also a working mom, is in her home gym, cycling in her own personal spin class, bright lights showcasing her athletic body.

Allison's longevity genes—mTOR, for example—are regulated by her high-intensity exercise. She's activating her brain-derived neurotrophic factor (BDNF), which is like Miracle-Gro for the brain's ability to focus and perform executive functions. Allison didn't inherit the best genes; obesity and diabetes run in her family. Despite her heredity, she knows the secret work-around: discipline, willpower, and smart habits, especially when it comes to sleep, exercise, and organic plant-based food.

Allison is forty-one years young and five feet six inches tall, and she's been the same weight since high school: a hundred and twenty-five pounds. She gained twenty-seven pounds during each of her three pregnancies, but those pounds melted off about five minutes after she gave birth (okay, maybe it was more like eight weeks, but you get the picture). She is in bed by nine thirty, regardless of her husband's late-night activity. Allison is motivated to adjust her lifestyle to turn off her bad genes and turn on her longevity genes. Allison looks healthy and strong, but she earns it the hard way. You can too, even if you're not a morning person or hate to sweat (or both).

Tragically, over the past century, Americans have engineered movement out of their lives. Today, we sit more and move less than ever. As much as we know exercise is good for us, only about 20 percent of us do it regularly. On top of that, 70 percent of Americans work at desk jobs. Between commuting to and from work, sitting at a desk, and then coming home to

Flap Your Arms

Music conductors live longer than members of any other profession. Most conductors live into their eighties, nineties, and older. They don't live stress-free on a mountainous island like the Icarians do, gathering wild greens and brewing tea. Like Blanche Honegger Moyse, who died recently at a hundred and two, they travel the world and stay up late for rehearsals and performances. As they flap their arms during conducting, they demonstrate relentless physical stamina, charisma, and passion in their performance.

Leopold Anthony Stokowski (1882–1977) is perhaps best known as the conductor of Disney's *Fantasia*, but what makes him stand out in my mind is that he was conducting right up to his death, at the age of ninety-five. Ah, Maestro!

Other professions that fare well in longevity include folks who sit very little: archeologists and astronauts, clergy, teachers, and medical doctors. Some of the worst careers for longevity include watchmaking and working in the textile industry, perhaps because of prolonged sitting and exposure to toxic chemicals.[1]

sit at the dinner table or in front of the television, the average American spends eight hours per day sitting.[2] No wonder obesity is spreading in an epidemic fashion.

I'll cut to the chase: sitting accelerates aging. If you're a woman, that translates as follows: Sitting six hours or more per day increases your risk of cancer by 10 percent and your risk of early death by 34 percent. (Men who sat six hours per day in the same study were 17 percent more likely to die early compared with the less-frequent sitters.)[3] Exercise partially mitigates the damage; women who sit a lot *and* don't exercise much are nearly twice as likely to die early than those who sit fewer than three hours per day and are physically active.

Why It Matters

Yes, sitting is the new smoking because of how it increases your risk of diabetes and heart disease—but it can also make your belly look fat by tightening your poor hip flexors and increasing your waist circumference.[4] Here's how that works: When you slump in a chair, abdominal muscles go slack while lower back muscles tighten, leading to a swayback. The sway in your low back pushes out the abdomen, and your hip flexors are too tight to pull the belly back into the core. The result is a bulging belly. In the next chapter, you'll learn a basic move that will get your hip flexors working properly again, tucking your belly back behind your abdominal muscles where it belongs.

Issues of vanity aside, when you sit too much, you trigger a cascade of bad outcomes.

- Weak bones—too much sitting is one of the reasons that the incidence of osteopenia (thinning bones) and osteoporosis is rising

- Organ damage to heart, pancreas, and colon

- Muscle decline[5]

- Hormone problems (just one day of prolonged sitting lowers your insulin response)

- Bad back (disc compression, inflexible spine)

- Poor circulation in legs (varicose veins)

- Overall, prolonged sitting increases the risk of diabetes by 112 percent, the risk of cardiovascular events by 147 percent, the risk of cancer by 29 percent, and the risk of all-cause mortality by 50 percent.[6]

Even if you're already exercising an hour each day, your workout can't offset all the damage of excessive sitting; you must sit less.[7] Happily, even allocating ten minutes from sedentary time to moderate-to-vigorous activity can significantly reduce waist circumference, not to mention provide a host of other benefits.[8] Let's counteract the harm of sitting and continue your epic mission to outsmart aging and early death.

Beat Early Death

Overall, physical activity reduces the risk of early death by 30 percent in men and 42 to 48 percent in women.[9] In every one of thirty-eight studies, physically active women lived longer than inactive women.[10] Moderate exercise is enough to put off premature death; even though high-intensity exercise may be better for short-term benefits, in the long term, they are both good for you.

- Is it ever too late? If you've waited until age sixty-five or even older, is it too late to get fit? Women who became more active after age sixty-five still have a mortality benefit.[11] So even if you're seventy and you've never exercised, it's not too late to stave off early death. Hooray!

- What if I'm fat? Even if you're overweight, exercise reduces your risk of dying prematurely. Although your best bet is to be lean and active, the worst-case scenario is to be obese *and* inactive. Plus, exercise will help you get lean and lose excess body fat.

The Motive to Move

One of the benefits of exercise is that you can convert ugly white fat, concentrated in your belly and subcutaneous (under-the-skin) tissue, into something close to the virtuous brown fat that burns calories and generates heat and is located mostly in your neck and shoulders. White fat puts you at greater risk for diabetes and heart disease. Exercise shrinks the size of your white fat, and, ultimately, turns it beige (a mix of the two).[12]

Exercise also alters the expression of thousands of genes. Specific movements activate the longevity genes. Genes such as ADRB2 regulate your weight changes in response to exercise. When you exercise more than three hours a week, you'll turn on APOA1, the gene involved in HDL, or good cholesterol, production.

One of the good variations I have is in a group of genes—LPL, PPARD, and LIPC—that give me enhanced benefits from endurance exercise, so I need longer walks and hikes at moderate intensity to fully leverage my DNA. Auspiciously, my fitness buddy has the same endurance genes, so we're ideal partners. No matter your genetic makeup, the exercises in this chapter will turn on and off the most critical factors that protect against aging, keeping you literally young at heart.

Top Ten Reasons Women Don't Exercise (and the Counterarguments)

So if we all know exercise is good for the body, why is it that 88 percent of Americans don't exercise? Perhaps many are like me—I don't love it. By nature, I'm apathetic about exercise and always have something better to do. Exercise, especially the tough stuff like CrossFit or any other high-intensity interval training, just plain exhausts me. Some people respond well to exercise—they get ripped and lean—but I'm not one of them. I work out hard, I mean *really hard*, six days per week, and looking at my body, you can hardly tell the difference between me and a couch potato. So why bother?

Here are the top ten reasons I hear for why women don't exercise, and the counterarguments that get me to lace up the running shoes or head to barre class.

TOP 10
REASONS WOMEN DON'T EXERCISE
(AND THE COUNTERARGUMENTS)

❶ "NO TIME."

℞ Squeeze a little extra activity into your day, because everything counts. Try smaller bits of exercise, like pacing while you talk on the phone or taking a 2-minute dance break while you brew your coffee. Science geeks call it "Non-Exercise Activity Thermogenesis" (NEAT), or all the ways you burn extra calories through fidgeting, pacing, parking farther away so you walk more—it adds up!

❷ "EASILY BORED."

℞ Vary the plan. Try the menu approach, and pick the "dish" that suits you in the moment: Zumba, barre class, chi running, CrossFit, qi gong. Find exercise that feels more like play and less like something you have to do for the sake of exercise.

❸ "NOT A MORNING PERSON."

℞ Keep your workout short, do it at lunch, or meet a friend in the afternoon or evening.

❹ "NEGLECTING FAMILY."

℞ Take them with you. Give your kids incentives to hike with you (have you heard of geocaching?), or drag your partner to yoga or dance class.

❺ "TOO TIRED AFTER WORK."

℞ After 5 P.M., when my willpower is gone, it helps me to create "if/then" statements with alternatives, such as "If I'm too tired, then I'll go on a sunset stroll with my husband."

❻ "CLUMSY, INEPT, IGNORANT."

℞ Take a class, or work out with a trainer. Don't make it worse by showing up for an advanced hip-hop class. Trust me, I know from personal experience that barre class is great for non-athletes. If you don't know anything about strength training, stretching, and the like, go to a beginner's class or orientation.

❼ "TOO SICK (OR INJURED)."

℞ You may need to work with an expert who can help you adapt exercise to your individual needs. Focus on setting a learning goal, not a performance goal. Don't do too much, too soon. Cut yourself some slack.

❽ "NO STICK (OR FOLLOW-THROUGH)."

℞ Studies show that when you "start out whole hog" and lose steam, it's better to focus on immediate rewards, such as more energy, less stress, and better mood. Accountability, working out with a buddy or family, also doubles to triples your results. For instance, I run every Sunday with my friend Jo, rain or shine.

❾ "SWEAT IS DISGUSTING."

℞ Two words: moisture wicking. Look for other buzzwords such as "the silver bullet for stink" or "stretchy ventilation." Get some high-tech workout clothing with breathability, and choose one of the many exercise methods that make you sweat (or notice it) less: swimming, brisk walking, and barre class.

❿ "LAZY."

℞ You need accountability, like a workout partner or a trainer, and to schedule exercise when you're most energized. You may also need to fake interest until it becomes real, the Twelve Step adage of "Act as if . . ."

A Window into Your Longevity: Resting Heart Rate

In addition to the more obvious benefits of exercise, movement affects your heart rate, and heart rate affects your healthspan. Recently, my husband, David, raced up a mountain at a spa north of San Diego, California, leaving a group of twenty other guests and me in his dust. It wasn't even 7:00 A.M. My husband has the sprinter's gene, which codes for explosive power that many Olympic athletes have. I do not have the sprinter's gene; I have the forget-this-let's-go-have-breakfast gene. Why work so hard while on vacation? Isn't the point to chill out when you have the chance and remember why you got married? How about a pot of green tea, followed by reading the newspaper and getting a massage?

While my husband was sprinting ahead, I was panting and hiking up the steep slope as fast as my heart would let me. I was surrounded by interesting women, including the owner of a biodynamic winery in Sonoma and another who was a soap-opera actress from Brazil, but I was breathing way too hard to talk.

David is a lifelong athlete. I'm not. He has a lovely physiological adaptation called athletic bradycardia, which is a fancy way of saying that his heart beats very slowly at rest because he is forever fit with indelible muscle memory. Athletic bradycardia, or athletic heart syndrome, as it is sometimes known, is common in athletes who routinely exercise more than one hour per day. As mentioned, even when my husband is not in good aerobic condition, his heart retains the memory of his years on the football field and the track. We've even measured it; when my husband is in bed at night and while he's sleeping, his heart rate hovers in the mid-40s. My resting heart rate is what's considered normal, consistently in the 60s, or the 70s when I'm stressed.

David is fifty-six, and so his maximum heart rate is estimated to be about 164 beats per minute. That means his heart can go up and down a broader range so he can meet the challenge in front of him, like riding his bike straight up Mount Diablo, where his heart rate is about 160 to 165 and he feels strong. By contrast, when I get near my maximum heart rate, I huff and puff and feel like I'm going to die. (Maximum heart rate is simply a guide; some people, like my husband or a twenty-five-year-old sprinter, can exercise and go all out, even above their calculated maximum, without problems and feel fine. Others who are less fit develop chest pain or fatigue. Allow how you feel to guide you, and be careful to track how you feel at various heart rates while exercising.)

Not surprisingly, resting heart rate is a marker of longevity. It makes sense; the more efficient your heart is (that is, the slower your resting heart rate), the longer it keeps ticking. There's an upper limit on how many times your ticker will tick.

That day at the spa, David chased a very fit thirty-year-old female trainer up the mountain. An hour later at breakfast, David flashed me an endorphin-soaked grin. I was still out of breath when he remarked "The leading trainer has a resting heart rate of forty-two, and every time I almost caught up to her, she would find a faster gear and race ahead of me. She wouldn't talk to me or tell me her strategy. She trained her husband for the Navy SEALs. It was awesome!" He paused a beat, perhaps wondering what took me so long to complete the hike, then whispered: "Next year, I'm totally training in advance so I can pass her."

Apparently, I married an aspiring Navy SEAL. I dug into my spinach omelet and thought of Sylvia from my barre class—resting heart rate: fifty.

Many scientists believe that humans get only a fixed number of heartbeats over a lifetime, so efficiency means longevity. (The actual number of heartbeats per human is estimated to be around 2.2 billion.) One investigator, Dr. Herbert J. Levine, found that across the animal kingdom, there's a direct relationship between heart rate and life span: the faster the heart rate at rest, the shorter the life; the slower the rate, the longer animals lived.[13] The results were enough for him to weave wit and wisdom and develop the adage "You're probably born with a certain number of heartbeats. Don't use them up too fast."

The idea is that fitness conditioning allows your heart to get more blood to your muscles more efficiently and at a lower heart rate. Indeed, *cardiac slowing at rest* may be one of the ways that exercise extends your life. You'll learn more about how to calculate your target heart rate and the optimal range for you later in this chapter.

Science of Week 3: Move

These days, everyone knows exercise burns away tension, builds muscles, conditions the heart and lungs, and cranks up the happy-brain chemicals known as endorphins, but there's so much more to it than that. In my experience, many people come to exercise for all the wrong reasons (often involving magical thinking about weight loss), but stay for the right reason: how it makes them feel afterward—the improved energy, mood (from endorphins), feeling better about themselves for doing it, and brain function. Maybe you hate exercise, but you love your health and are smart enough to figure out that you need exercise to be your best version of yourself. That's more than just a theory; it's been proven scientifically, especially for people who find forms of exercise that they actually enjoy.

If you just want to know what to do and not get lost in the science, feel free to skip ahead. Go straight to the protocol to get started on changing the way you move and exercise to lengthen your healthspan.

How Much Exercise Is Best?

Currently, physical activity guidelines suggest that you get a hundred and fifty minutes per week of moderate physical activity, seventy-five minutes per week of strenuous activity, or an equivalent combination of both.[14] But dose matters, and more isn't necessarily better.

In college, I frequently changed the type of exercise I did. I ran aggressively for four miles, same pace, four times per week, then grew bored and switched to Kathy Smith fitness videos for a few months, then switched to weight training for an hour four times per week, with maybe a weekly game of squash. Then I became a rower, again following my rule of fours: one hour of hard rowing four times per week. My best friend's mother had a more consistent schedule: she jogged two miles in twenty to thirty minutes five times a week. I found the simplicity of her steady regimen refreshing, and it turns out that she was onto something important.

When it comes to exercise and healthspan, moderate may be best because there's a point of diminishing returns. Even the form of exercise might matter. The relationship curve seems to be U-shaped, meaning sedentary folks *and* strenuous exercisers fare the worst, and moderate exercisers fare the best, at least for joggers.[15] For my friend's mother, she lowered her mortality by 44 percent with her two-mile, five-times-a-week jog. Two new studies suggest that when it comes to jogging and running, doing about one to two hours per week, ideally broken into a few twenty- to thirty-minute sessions of two to three miles, seems to be best.[16] Higher doses of extreme exercise such as marathons, ultramarathons, and full-distance triathlons may cause cardiotoxicity (harm to the heart).[17]

Studies of the biological timekeepers in your cells, telomeres—the caps on your chromosomes that stabilize your genome—show that moderate levels of activity seem to protect telomere length best.[18] Furthermore, both moderate- and high-intensity exercise at a dose of two to four hours per week was found to benefit women's telomere length.[19] There was no additional benefit to telomere length for more exercise, so moderate levels of activity are sufficient and that's what I advise.

How do you dial in the right dose? Mortality studies suggest one to two hours per week of moderate activity, and telomere studies show a benefit at two to four hours per week. So exercise dose should be between two and four hours a week for the average person. If you're not average, adjust based on how it makes you feel. My husband feels fantastic riding his bike about ten hours per week (one hour on a stationary bike three days a week, and two multi-hour rides on the weekend) and more when he's training for his next century race. That's the right dose for him. For me, yoga three times per week, one barre class, daily walks, plus my run with Jo is about right.

What Type of Exercise?

Overall and for everyone, it depends, although a balance of cardio and strength is important. In the average woman and based on limited evidence by measuring telomere length, yoga, calisthenics, and any aerobic activity may be best.[20] Brisk walking, swimming, Pilates, and bicycling are also good, although results didn't reach statistical significance in the largest study to date. In functional medicine, we adapt the prescription to the individual rather than apply one-size-fits-all rules. While a regimen of running short distances five times a week worked for my friend's mom, you may be best served by a cardio Pilates class four times per week or hiking in nature.

I always get asked about walking and bone density in postmenopausal women, so here's the skinny: walking improves one small part of the hip, but not other parts of your skeleton.[21] So it's a good idea to mix strength training with brisk walking for optimal bone density. One meta-analysis found that women who practiced combined resistance training (high-impact aerobics together with resistance exercise) showed the greatest improvement in bone density at the hip and spine.[22]

Women, who are more prone to osteoporosis than men, can help protect themselves with regular weight-bearing exercise (two to three times per week with handheld weights, body resistance, or weight machines) or yoga—yes, yoga counts as strength building (although the data are mixed in postmenopausal women regarding bone density, yoga slows down bone resorption and has other health benefits[23]). Regular strength training increases your bone density, helping counter both hormonal and metabolic changes that can cause osteoporosis. One study recruited a group of people seventy and older to perform resistance training twice per week for six months. These poor folks submitted to a muscle biopsy at the beginning and again after six months of strength training, so please take their important results to heart so that their service was not in vain. Ever had a muscle biopsy? Looking at the results of the second muscle biopsy, researchers found that 596 genes had reverted to a much younger state, giving a portrait of youthful vigor similar to healthy, active controls in their twenties who'd undergone the same muscle biopsies.[24]

Increased muscle mass from strength training also keeps the metabolism revved up, which improves the body's ability to burn fat. That's why strength training is crucial as you age. If you want to burn belly or visceral fat and take on the muscle factor from the introduction, drop into down dog. In a study published in the journal *Menopause*, sixteen healthy but obese postmenopausal women with an average age of fifty-five (and an average body-fat composition of at least 36 percent) were randomly assigned to do yoga three times a week for one hour or to a control group.[25] The control group performed no exercise. After sixteen weeks, the yoga group achieved the following stellar results:

- Lower body weight, due to a significant drop in body fat (especially visceral fat, which ages you) and an increase in lean body mass (which increases healthspan)

- Smaller waist circumference (waists greater than thirty-five inches in women are one of the markers of the dreaded metabolic syndrome)

- Better adiponectin level (the hormone that burns fat and tends to run low if you have a variant of the gene ADIPOQ)

- Improved cholesterol pattern (higher HDL, or good cholesterol, and lower total cholesterol, LDL, and triglycerides)

- Lower blood pressure, insulin, glucose, and insulin resistance

Another strategy for burning fat is to bump up the intensity of your aerobic routine—not continuously but intermittently. In a study from 2008, twenty-seven overweight women with an average age of fifty-one and an average BMI of 34 who performed bouts of high-intensity aerobic exercise reduced their waistlines and visceral fat more than women who did not exercise or did only low-intensity aerobic activity. The higher-intensity exercise modulates the mTOR gene and gets your body to produce more growth hormone, which helps to reduce visceral fat. Interval training or burst exercise stimulates your longevity genes. When properly regulated, this gene improves the proteins that allow your muscles to contract and move hard. The mTOR gene in the brain improves learning and memory; mTOR in the heart helps remodel the heart muscle so it can beat more efficiently with a larger volume at a slower resting heart rate.[26] The mTOR effect on muscle (skeletal and heart) contraction is similar regardless of gender or age, so it's never too late to jump in.[27]

After intermittent fasting, burst training with any type of exercise may be best for longevity. Studies show that when you exercise intensely for five to thirty minutes while still fasting, you are most likely to reset mTOR. Intermittent fasting turns off mTOR, which can become dys-regulated and contribute to accelerated aging, cancer, and Alzheimer's disease. In addition to regulating mTOR, exercise makes your receptors more sensitive to insulin, gets your blood sugar under control, and helps you build more muscle. The combination of intermittent fasting followed by high-intensity exercise will boost growth hormone and a hormone called irisin, inducing white fat into behaving like brown fat and building muscle.

Here are a few ideas to help you do that: walk briskly for two minutes alternating with jogging for two minutes (for a total of twenty minutes) or bicycle fast for bursts of one minute alternating with a slower pace for one minute for a total of thirty minutes.

The idea is to overload your muscles' mechanoreceptors—the sensing devices for how hard and fast you're working the muscle. The key with your exercise is targeted, smart overload followed by adequate and active rest. You don't want to exercise at a high intensity for more than forty-five minutes because that increases oxidative stress and cortisol, the wear-and-tear hormone that hastens aging. (If you're an endurance athlete, you'll need additional measures to counteract oxidative stress and high levels of cortisol.)

When it comes to genes and the types of exercise that activate them, the effect is gigantic: exercise causes methylation changes at eighteen thousand sites on 7,663 genes.[28] My goal is for you to have a general sense of the gene-exercise interaction, not for you to memorize every gene (see the appendix for a partial list of key genes). Regardless of your genetic makeup, exercise turns on key longevity-promoting genes and turns off life-span-shortening genes. I won't bore you with the full list of those genes, but here are five highlights of exercise.

1. Turns off the Fatso (FTO) gene. Allison, the friend I described at the beginning of this chapter, and I went on a girls' trip for her fortieth birthday. We exercised daily, which is way more fun with girlfriends, because we both want to turn off the Fatso gene with vigorous exercise.[29] Indeed, keeping your weight down and your body fat in check may abolish the adverse effect of FTO by helping to normalize leptin, the satiety hormone.[30]

2. Raises insulin sensitivity and good cholesterol. Fortunately, we have the option to turn off the diabetes genes with exercise.[31] Researchers have learned that active women with the same variant of LIPC may improve good-cholesterol (HDL) levels and have fewer heart attacks than inactive women with this gene variant.[32]

3. Improves methylation. Methylation improves with even a single session of exercise. Greater intensity in exercise begets greater changes in methylation and other epigenetic patterns and results in a leaner you by making your muscles soak up blood sugar better and get stronger.[33] The more consistently you exercise, the more your cells adapt to help you maintain blood sugar, increase muscle mass, weather the aging process, and stay younger.[34]

Green Walks

Getting out in nature has benefits; a new study confirms the positive brain changes that occur. Walking in nature makes you more attentive and happy than walking the same amount of time close to heavy traffic or in an enclosed environment.[35] Nature makes you less likely to brood (defined as morbid rumination). Gregory Bratman at Stanford recently found out why. Apparently, a part of the human brain, called the subgenual prefrontal cortex, is the center of brooding. Walking in nature for ninety minutes quiets this part of the brain.[36] Many studies have documented that city dwellers with less access to green space have more psychological problems: anxiety, post-traumatic stress disorder, depression, and brooding.[37]

A Walk a Day Keeps Breast Cancer Away

Overall, exercise consistently reduces breast cancer risk by 12 to 60 percent.[38] After menopause, women who walk daily have a lower risk of breast cancer compared with sedentary controls.[39] Since one in eight women develops breast cancer over her lifetime, a walk a day is an easy decision. One hour per day of walking lowers risk by 14 percent.[40] You can even prevent precancer of the breast.[41]

4. Lowers blood pressure. High blood pressure runs in my family, but I inherited the solution: exercise. The gene EDN1 codes for endothelin-1, a potent constrictor of blood vessels. My variant of gene EDN1 puts me at a greater risk of high blood pressure if I'm inactive, and I'm not alone—21 percent of people of European descent, 41 percent of South Asians, 19 percent of Africans, and 14 percent of Latinos have the same variant.[42] As long as I keep exercising and following my Younger protocol, I have normal blood pressure.[43]

5. Allows wise choices. I've been reviewing patients' genomes for a long time, and I've never seen a person who wouldn't benefit from mid-distance endurance exercise. Here's what my friend Jo and I do every Sunday: we sprint/walk for sixty minutes, then follow that with a ninety-minute yoga class (it's taken years to train our husbands to accommodate this extravagance). The run/walk allows us to catch up and coach each other. The yoga lengthens our tight hip flexors. After two and a half hours, we are both much better wives and moms!

Benefits Galore

In the past twenty years, we've learned important and sometimes counterintuitive things about exercise in women. Here are the studies showing the areas that benefit from exercise.

Cognition. Exercise keeps you smarter longer. In fact, sleep and exercise are your two best weapons against Alzheimer's disease (covered more in chapter 11). Women sixty-five years or older in the highest quartile of fitness are less likely to develop cognitive decline than women who are not fit.[44] Long-term regular and higher levels of activity were associated with better cognitive performance and less cognitive decline in older women.[45] Takeaway message: walk more to keep your wits about you.

Here are a few theories about how exercise maintains your cognition:

- Physical activity sustains the brain's vascular health—that translates as better blood flow and oxygen supply so you can think better.

- Evidence shows a relationship between insulin and the Alzheimer's-related beta-amyloid plaques. It is possible that aerobic activity benefits insulin resistance and glucose intolerance, leading to fewer plaques.

- Fitness may directly preserve the structure and growth of your brain as you age.

Autonomic nervous system. A section of your vast nervous system is responsible for certain automatic body functions, such as heart rate, breathing, and digestion. Half of the autonomic nervous system is involved in "fight, flight, or freeze" (hereafter shortened to fight-or-flight). The other half is the parasympathetic nervous system, involved in "rest and digest." Healthy longevity may depend in particular on maintenance of parasympathetic tone, which tends to fall in the eighth decade.[46] As mentioned, elite athletes have higher tone in their parasympathetic nervous systems, resulting in lower heart rates. You can easily measure the function of your autonomic nervous system by checking your resting heart rate, your HRV (meaning the time between each heartbeat), or your heart-rate recovery after exercise. More to come on this topic in the following pages.

Well-being. While it may seem vague as an outcome of exercise, one study quantified well-being and the effect of leisure-time activity over thirty-two years.[47] When women either increased or decreased their physical activity, a corresponding change occurred in their well-being. Inactive women reported four- to sevenfold lower well-being.

Firmer skin. Some of the genes affected by exercise even involve your skin. Exercise keeps your skin young and may turn around sagging and other forms of skin aging even when you start to exercise later in life. As you probably know from watching the mirror, skin changes as you age, resulting in wrinkles, translucency, crow's-feet, and sagging. Mostly these changes occur as age affects the different layers of the skin, independent of other aging factors like sun damage. When you're around age forty, the outermost layer of skin, called the stratum corneum, thickens and becomes more dry, flaky, and dense. With age, the epidermis loses some collagen but remains relatively unscathed, while the underlying and

Surprise Results of One-Legged Bicycling

You can change the expression of thousands of genes by bicycling. Researchers at the Karolinska Institute in Sweden designed a brilliant study to look at the epigenetic changes of exercise. They tracked twenty-three young men and women before and after three months of bicycling with one leg only.

The one-legged cyclists were to pedal at a moderate cadence four times per week for three months. In other words, each person became his or her own control: one leg was exercised, one was not, and both legs were biopsied. (I can't quite imagine what the subjects looked like after three months with one leg beefy and the other neglected; they must have been paid well!) More than five thousand sites on the DNA from the exercised leg had new methylation patterns. Gene expression was substantially increased or changed in *thousands* of muscle-cell genes, including those that controlled insulin sensitivity, energy metabolism, and inflammation. Overall, these changes made the muscles more healthy and functional via a relatively short amount of endurance training. For you, that means less belly fat and more power and energy available throughout the day.

innermost layer of skin, called the dermis, starts to thin. But not in everyone. If you exercise, your outer layer of skin doesn't thicken as early, and your inner layer of skin doesn't become thinner!

Mark Tarnopolsky, a professor of sports medicine at McMaster University, took a group of sedentary people aged twenty to eighty-six and got them to exercise. The subjects exercised for just thirty minutes twice a week, jogging or cycling at a moderate-to-vigorous pace (65 percent of their max heart rate), and at the end of three months, researchers found that the skin of the older subjects looked like the twenty- to forty-year-olds' in the same study.[48]

So what's responsible for the reversal in age-related skin changes? Certain myokines, which are proteins released by working muscles, enter the blood and increase before and after exercise. Your skin uses myokines: the more myokines provided to your skin, the younger it stays.

Better sleep. Exercise heats your body, so later when your temperature cools down, it may help to induce high-quality sleep. A number of small studies show that morning exercise is best, ideally before 1:00 P.M. Other data of larger populations suggest that timing may not matter; just find a way to fit it in![49]

More relevant to aging, in those in middle and older age with sleep problems, training programs of ten to sixteen weeks moderately improve sleep. Overall, the exercisers fell asleep faster and needed less medication.[50] Specific training programs included the following:

- Moderate-intensity aerobic exercise on a treadmill or bicycle for thirty to sixty minutes three to five times a week at 50 to 75 percent of maximum heart rate

- High-intensity resistance training for sixty minutes three times per week

- Tai chi for forty minutes three times per week

Now that you know some of the benefits of exercise and all the ways you can put your body in action, it's time to move smarter, sleep deeper, and reduce inflammation starting this week. You have no excuses now.

Tai Chi

Tai chi deserves more attention. It's one of the best exercises for people with trouble sleeping. Tai chi, a Chinese martial art described as meditation in motion that promotes serenity through gentle and flowing movements, may enhance your sleep quality, sleep efficiency, and awakenings, and it may add about forty-eight minutes to your time of slumber.[51] In breast-cancer survivors with insomnia, tai chi reduced inflammation.[52]

Protocol for Week 3: Move

Starting in the preparation phase (chapter 4), you began exercising twenty to thirty minutes four days per week. It's time to refine your goals, and it's never too late to pick up exercise. Strength training improves longevity even in frail people in their nineties![53] After eight weeks of high-intensity training, strength increased 174 percent and midthigh muscle area grew 9 percent. *What?* They even walked almost 50 percent faster.

When you're deciding what type of exercise to choose, the most important factor is that you like it. Try to get outdoors if you can.

Before you start, calculate your maximum heart rate (MHR) plus your heart-rate training zones. If you type *target heart rate calculator* into a search engine, you will find several options for calculating your heart rate at different intensity levels (or percentages). Here are the steps to do it yourself:

- Take 220 minus your age.

- Depending on your goals, you will train in the green (70 to 80 percent of your maximum heart rate), orange (80 to 90 percent of your maximum heart rate), or red zone (90 to 100 percent of maximum heart rate). For me, that's 220 − 49 = 171 maximum heart rate. So my zones are

    − Green zone (moderate) is a heart rate of 117 to 136 beats per minute (bpm).

- Orange zone (vigorous) is 136 to 154 bpm.

- Red zone (maximum) is 154 to 171 bpm.

Now it's your turn:

---

220 – _____ (your age) = _____ (max heart rate [MHR])

Green zone: 0.7 * MHR _____ – 0.8 * MHR _____

Orange zone: 0.8 * MHR _____ – 0.9 * MHR _____

Red zone: 0.9 * MHR _____ –   MHR _____

Exercise in the moderate (green) zone to improve aerobic fitness, and the vigorous (orange) zone to improve performance capacity and postexercise oxygen consumption, so that you burn calories after your workout. Up it to the orange or red zone briefly (thirty seconds to a maximum of three minutes) for high-intensity interval training—it will help you increase performance and speed. Exercise at the moderate to vigorous level has been shown to increase longevity.[54]

HEART RATE TRAINING ZONES FOR EXERCISE

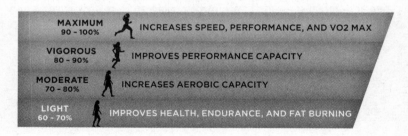

MAXIMUM 90 – 100%	INCREASES SPEED, PERFORMANCE, AND VO2 MAX
VIGOROUS 80 – 90%	IMPROVES PERFORMANCE CAPACITY
MODERATE 70 – 80%	INCREASES AEROBIC CAPACITY
LIGHT 60 – 70%	IMPROVES HEALTH, ENDURANCE, AND FAT BURNING

Here is your template for week 3. For the next seven days, follow these guidelines as closely as possible and notice the changes as exercise interacts with your genes and hormone levels.

Basic Rituals

- Sit less. Even if you have a desk job, get up at least once every forty-five minutes. Set a timer on your smartphone or wear a tracker that measures your standing time. Do a one-minute burst of enthusiastic dancing after sitting for an hour. Invest in a stand-up or treadmill desk, and use it daily (I've walked more than two thousand miles on mine while writing this book).

- Move more. Aim to find bursts of movement that fold into your natural rhythm. Practice fifty heel lifts while you chat on the phone or stand in line at the grocery store. Perform twelve push-ups after going to the bathroom. The point is to find moments of movement rather than only forced discipline that's devoid of pleasure. Add one to five minutes of new movement to your routine each day during this week.

- Burst-train two to three times per week, an exercise where you focus on fast-twitch muscle fibers. Cavemen and -women tended to exercise in bursts: a quick run to the river to fetch water and carry a bucket back to the tribe, a jog with a sick infant to a neighbor's dwelling for help. Our bodies perform well with burst training, and then recover at a moderate intensity for one to three minutes. Protocols vary; use one that makes the most sense for you. Burst training can be applied to cardio exercise (e.g., intermittently sprinting on a trail alternating with a jog) or weight lifting (lifting a weight, such as with a biceps curl, as many times as you can with good form for one minute, followed by one minute of rest). Other examples:

  - Walking three minutes fast (approximately 6 or 7 on an exertion scale from 1 to 10, or the green zone of 70 to 80 percent of your maximal exertion), then alternating with three minutes at a normal pace[55]

  - Chi running with sprint intervals, or regular running with 30-second sprints

- High-intensity interval training with weights or cardio (stationary bike, elliptical, treadmill), alternating two to three minutes at a moderate pace with one to two minutes at maximal pace for you

• After burst training, get a recovery drink. It increases muscle mass and keeps mTOR turned off. This is only for people who perform burst training (at least four to five bursts per session) or vigorous training of at least thirty minutes' duration. What's proven to work is a combination of macronutrients high in protein and carbohydrates, even in older folks. But drink it within forty-five minutes of your workout; immediately after your workout is ideal.[56] Avoid sugar. The best formula is somewhere between 10 to 40 grams protein (I suggest 20 grams for the average woman), 7 grams or more of carbohydrate (I suggest 10 to 20 grams for women), and up to 3 grams of fat. See Resources for recommendations.

• Keep up the rituals from weeks 1 and 2, including getting to bed before ten and sleeping seven to eight and a half hours.

• Schedule and take sufficient time for recovery. I used to think *recovery* meant when I wasn't sore from yoga or running. Or it was twenty-four hours after my last fitness spree. Or maybe recovery was when I could exercise again without feeling like crap. Turns out that I didn't know the first thing about recovery, because it's more about galvanizing the full arsenal of repair mechanisms in your body: stitching together microtears in your muscles, ironing out the fascia when it gets jangled, reinvigorating mitochondria so you're brimming with energy rather than feeling worn down or burned out. Adequate recovery keeps your hormone profile in balance so that your adrenals don't get fried and take your sex hormones and thyroid down with them. The official definition of *recovery* is your ability to repair tissues damaged during exercise, rebuild muscles, provide functional restoration of the body such that you prevent injury, rejuvenate emotionally and psychologically, and feel prepared to meet or exceed performance the next time.

Previously, I'd chronically limit my recovery, and I wonder if the same is true for you. If you exercise five days per week, then at its simplest, recovery means twenty-four hours between bouts of exercise and two rest days. If you exercise four days per week, you take three rest days. For me, my weekends are my harder exercise days, and Mondays and Fridays are my rest days.

Recovery allows you to heal from oxidative stress, which you may or may not feel as fatigue and muscle soreness. But recovery runs deeper; in a larger sense, it's about paying attention to the messages of your cells, your inner voice, and not letting ego run the show. My ego tells me to overexercise and under-recover, which is a recipe for injury, spasm, and weak mitochondria. Don't let that happen to you. Recovery is also about tuning into the messages your body is sending you—the ache in your left sacroiliac joint (for me, a sure sign that I'm heading toward spasm and need to activate my release rituals, covered in chapter 8) or the twinge in your right knee. Ironically, I taught myself to ignore those signals during medical residency when self-care came last, but more recently I've been learning to hear and feel those sacred messages from my body in my recovery.

If you want to be a recovery goddess who turns on and off the right genes, learn about heart rate variability training in the Advanced Projects section. For many fitness enthusiasts, measuring heart rate variability is the best, easiest, and most objective measure of whether the body is ready to train—it tells you whether your nervous system is prepared for another dose.[57]

Supplements

In addition to intermittent fasting and high-intensity interval training, taking branch-chain amino acids (BCAAs) may help you regulate mTOR activity. Drink BCAAs mixed with water or as an oral supplement during your high-intensity workouts. The typical dose is 3 to 8 grams/day. Your blend of branch-chain amino acids should contain leucine, isoleucine, and valine. Check Resources for brands that I recommend.

Advanced Projects

- Measure how long it takes you to run one mile. Use a treadmill or track. If you cannot run, aim for a jog, and, if you must, intermittently walk and jog. Based on two studies in Texas, you can predict your risk of heart problems in older age based on how fast you run a mile in middle age. Researchers examined fitness levels in 66,371 people and found that the time it takes a person to run a mile in her forties predicts that individual's heart health as reliably as cholesterol level or blood pressure.[58] An additional study deemed that a woman in her fifties who can run a mile in nine minutes or less has a high level of fitness. If it takes her ten and a half minutes to run a mile, she's at a moderate fitness level. If she's slower than twelve minutes, she's in low fitness. For now, simply record the time it takes you to complete one mile and aim to gradually improve it.

- Keep a journal for the week and record your exercise—maybe even including metabolic equivalents of task (METs). The purpose of this project is to provide accountability and establish the baseline from which you'll improve. *METs* can be used to compare relative intensity of various activities. It may also help you to use a tracker and record steps or heart rate. Determining METs is easiest on exercise equipment such an elliptical, treadmill, or stationary bike. One MET is equal to 1 kcal/kg/hour, so for the geeks, it is straightforward to calculate energy expenditure of different physical activities. The MET of sleeping is about 0.9; walking is about 3 to 4 METs, gardening is 5 METs, sex is approximately 6 METs, and running is 8.

  So, to estimate how many calories a woman weighing 150 pounds (68 kg) would burn during thirty minutes of gardening (5 METs), the calculation is calories burned = 5 METs × (68 kg) × 0.5 hour = 171 kcal (more commonly known as calories). You can use the formula above to estimate the energy expenditure for any activity as long as you know the METs. Aim for 7.5 MET-hours this week, or more.

- Determine your VO_2 max. When you work out, there's an upper limit to how hard you can breathe and the maximum amount of oxygen you can take in. This is your VO_2 max, and it drives oxygenated blood to your muscles so they can perform at their best. By measuring your maximum oxygen consumption (VO_2) and heart rate, you can calculate target heart rate zones based on your aerobic and anaerobic thresholds. Perhaps more important, you can establish a baseline for your VO_2 max so that you track it over time and prevent it from slipping. This test collects your physiological performance data during a graded exercise test while you're breathing into a mask. Many gyms provide this type of testing under the supervision of certified trainers.

- Measure your body composition, including total body fat mass and lean body mass. I like to use the Bod Pod. You sit in an enclosed chamber, shaped like a pod, while the air pressure is adjusted, and the air displacement of your body is used to calculate body composition, resting metabolic rate (the number of calories your body needs to support its basic functions), and total energy expenditure (the number of calories you need to make it through your day). The Bod Pod test costs about sixty to a hundred dollars.

- Heart rate variability measurement. Previously I mentioned that the balance between the sympathetic and parasympathetic halves of your nervous system was an important indicator of your aging process and ability to recover from exercise. Here's how it works: your sympathetic nervous system controls your heart rate via the release of epinephrine and norepinephrine, which increase HRV (the time between each heart beat). Your parasympathetic nervous system controls heart rate via acetylcholine, which decreases HRV. Many things influence HRV, including respiratory rate, blood pressure, temperature, stress level, and mind-set. If you're rested and recovered from exercise, there's a balanced state between the two parts of the nervous system, and you have a high level of HRV, ideally around 60 to 100 (the higher, the better). The full HRV scale is 0 to 100. Purchase a wireless heart-rate monitor and download an app to measure HRV to see if your nervous system is ready for your next fitness endeavor (see Resources).

- Walk or bike more (ditch the car). Be like Sylvia from chapter 4. While she doesn't even own a car, you don't have to go quite that far; you can choose to walk more, commute via public transportation, or ride your bike as part of this week's protocol. Numerous studies show that the citizens of regions with more active commuting or travel, such as Canada and the Netherlands, age slower, as indicated by a lower incidence of diabetes, high blood pressure, weight gain, and other cardiovascular risks.[59] Active commute and travel offers more psychological benefits and is more relaxing than driving.[60]

Your Daily Routine

The next page shows what a typical day might look like when you're utilizing the basic rituals from weeks 1 to 3. This is how my husband structures his weekdays. Personalize for yourself.

Recap: Benefits of Week 3

Exercise improves thousands of genes. Alter your genetic fingerprint with exercise. Sweat your way to better genes, a stronger brain, more satisfying sleep, and firmer skin.

Bottom Line

Truly, with all the data on exercise and how it benefits multiple genes, you no longer have an excuse to skip it. No one who's in middle age or

A TYPICAL DAY IN THE YOUNGER PROTOCOL: DAVID

7:00 A.M.	Wake up, test blood sugar, brush teeth with electric toothbrush, drink tea
7:15	Eat breakfast: usually gluten-free steel-cut oats, yogurt, berries.
	Take supplements: multivitamin, berberine, resveratrol, vitamin D, omega-3
	Review sleep tracker and resting heart rate
	Shower and shave, put on uniform (Lululemon top and pants, Hoka shoes)
8:00	Go to work
	Use stand-up desk, drink filtered water, maybe paint a canvas
Noon	Make and drink a green shake (see Recipes)
	Brush and floss
4:00 P.M.	Bike ride, spin class, or weights, stretching at gym
5:30	Sauna
6:30	Dinner, usually tons of vegetables, salad, clean protein, followed by dark chocolate!
7:30	Family time, catch up on news and e-mails
9:00	Put kids to bed
	Brush and floss
	Be with Sara and read
10:00	Go to bed, lights out

beyond can afford to leave out this crucial lifestyle element. Simply moving more than you do right now will jump-start the changes. As you add burst training in strength and cardio, your body, mind, and heart will thank you. Exercise will lengthen your healthspan and let you enjoy the most of the second half of life.

CHAPTER 8

Release
WEEK 4

Generally the rational brain can override the emotional brain, as long as our fears don't hijack us. But the moment we feel trapped, enraged, or rejected, we are vulnerable to activating old maps and to follow their directions. Change begins when we learn to "own" our emotional brains. That means learning to observe and tolerate the heartbreaking and gut-wrenching sensations that register misery and humiliation. Only after learning to bear what is going on inside can we start to befriend, rather than obliterate, the emotions that keep our maps fixed and immutable.

—Bessel van der Kolk, *The Body Keeps the Score: Brain, Mind, and Body in the Healing of Trauma*

I'm lying on my back in yoga class, miserable. We're on the third set of abdominal work, an exercise called the star-spangled pulse. *Ugh, I hate this one.* I soften my neck into the cradle of my hands, because as usual, my neck muscles are tight.

I inhale and press my upper sacrum into floor, then lift my tailbone. I raise my head and shoulders toward the ceiling on an exhale. The exercise continues; inhale, inwardly spiral right leg as it points to the ceiling; keep left foot on the floor. Exhale, pump right leg toward the ceiling three times. Repeat with the other leg. I should be developing a rhythm; instead, I feel awkward. Meanwhile, a constant voice inside my head increases in volume: *I hate abs! Am I ever going to get better at this? Why am I here? Coming to yoga in the morning is indulgent. I should be at the office getting real work done!*

My teacher calls out: "For those of you who are suffering and telling yourself the story that you can't wait for this to be over, it's time to shift your attitude. Because the way you do abs is the way you do life. And it can change for the better." I take a deep, full-lung-capacity breath as the words sink in. I get a bit more curious about what's happening in my belly and low back; the voice in my head quiets, my legs become more buoyant, and my tight spots (neck, sacrum, hip) finally release tension.

As the saying goes, pain is inevitable, but suffering is optional—even the suffering during abdominal work. That kind of exercise-related pain is crucial to unlock chronically restricted and tight pathways in the hips, psoas, quadratus lumborum, and sacroiliac joint. One of the reasons I practice yoga is that it makes my muscles and joints supple, rather than stiff, rigid, and even spastic. Other reasons: yoga is associated with longer telomeres, higher levels of nerve-growth factor, improved antioxidant status (and thus better defense against oxidative stress), reduced inflammatory gene expression, and increased healthspan.[1]

Yoga counteracts my natural tendencies. I'm an intensity junkie, addicted to high achievement. Even though I now know better and subscribe to a different paradigm, I still run myself ragged. I used to push myself to the extreme in all aspects of my life, from yoga (always trying the most advanced variation of every pose) to education, career, and even childbirth. I studied the Bradley Method and approached birth as an endurance sport, like a marathon. My ambition often left me isolated and in an endless cycle of pushing myself to the point of injury. It also exhausted my nervous system and sent my cortisol and blood-sugar levels into the red.

My constant drive created habitual holding patterns in my body. In Sanskrit these patterns are called *samskara*, and they kept me from experiencing the true gifts of yoga and life. Now I have the awareness to be in dialogue with my body and to stop powering through the hard parts of life, ignoring my body's cries for help. There may be ways that you're unwittingly cutting yourself off from gifts too, suffering as you resist what you know is good, reluctant to take the time to unwind and release.

I've learned how to back off, relax more, and release my chronic tightness. It's no longer about working harder. It's about *release* and about holding what I want and desire more lightly. It's about letting go of old compensatory patterns that no longer serve me.

Tension and habitual tightness in the body turn into the stiff joints and muscles that eventually lead to the reduced mobility of aging. In the ancient yoga tradition, activating *bandhas*, or yogic locks, in your body can reverse aging. It's beneficial to learn stretching techniques, *bandhas*, self-adjustments, and other methods to release tightness and increase your mobility in the long term. Here are a few of the most common genes that your release work will be talking to in this week's protocol:

- You'll turn off the genes that make you injury-prone, such as the Achilles gene (metallopeptidase 3 or MMP3).

- You'll turn off the genes that make you produce more oxidative stress in your mitochondria, such as uncoupling protein 2 (UCP2), a protein that weakens your muscles and contributes to the muscle factor discussed in the introduction.

- You'll turn off the genes that make your back unstable and cause you pain. The COMT gene is known as the "corporate warrior" gene, but if you have the met/met COMT variation of that gene, it makes you a worrier rather than warrior, and you may have problems clearing the brain chemicals associated with stress (such as adrenaline, noradrenaline, and dopamine). You may also have more trouble metabolizing estrogens. The COMT gene helps set your pain threshold, so people with the variant are more likely to be diagnosed with fibromyalgia (a chronic pain syndrome), low back pain, migraines, sciatica, or disability after lumbar disc herniation.[2] Given adequate release work, you'll reprogram your COMT gene expression if it tends to make you feel more pain.

- Another gene that modulates the experience of pain is BDNF, which you may recall codes for brain-derived neurotrophic factor.[3] In other words, both BDNF and COMT may cause a jump in pain perception, potentially making you more sensitive.

Of course, genetics is only a small part of the story—about 10 percent—when it comes to pain and stiffness. Environment is 90 percent responsible for how your genes are expressed; they are malleable and strongly influenced by all types of nutrients and anti-nutrients. Consider release an important nutrient in the mix. With adequate release, you'll improve your injury recovery time, clear stress from your body before it can result in disability, increase your range of motion, and boost your respiratory health.

In yoga and Pilates, there's a belief that you are only as young as your spine, or, as Joseph Pilates put it, you're only as young as your spine is flexible. Most of us don't move the spine enough, resulting in the decline of range of motion, structural balance, and mobility with age. It's the confining nature of modern life: driving, sitting, using computers, lounging in chairs and sofas, hauling around large purses. It makes you rigid, knocks you out of alignment, and creates poor posture. Most people don't notice it until they wake up one morning feeling stiff and like they've aged ten years overnight.

I wish I could report that ten minutes of focused yoga stretches could undo the damage done by your thirty-minute commute in your car and six hours of sitting at a desk. The truth is that we need to be wiser about how we craft the day—adding frequent stretches, realignment, and release—in order to repair the damage of confined postures and restrictive movements.

Why It Matters

If you look deep enough, you'll see that daily stress creates profound erosion in the mind, the spirit, and the body—in muscles, bones, ligaments, tendons, joints, the spinal column, around and inside cells. The body constantly works toward more stability and greater balance, or homeostasis—the relative equilibrium between independent physiological forces. In most of my patients, the balance tips in favor of excess wear and tear rather than toward growth and repair.

Most of us feel wear and tear as tightness, muscle fatigue, and joint pain. We see inflammation in the mirror as puffiness, belly fat and bloating, and tired eyes. Even if your muscles are relatively relaxed after a yoga class or massage, you still may chemically tighten your body with negative thoughts, the pressure to achieve, and reactions to the inevitable small stresses of daily life. It leads to low back and sacroiliac pain for me. For you, it may lead to disc problems, sciatica, hip degeneration, knee pain, scoliosis, immune system dysfunction, digestive issues, difficulty breathing deeply and freely, adrenal-gland depletion, emotional distress, and despair. The goal is to prevent the premature overuse of your body so that

Common Chronic Tight Spots, Head to Toe

- Jaw
- Neck
- Upper trapezius, shoulders
- Respiratory diaphragm
- Psoas and other hip flexors
- Low back
- Hips
- Sacroiliac joint
- Pelvic diaphragm
- Iliotibial band
- Quadriceps
- Hamstrings
- Achilles tendon (susceptibility to injury can be genetic)
- Feet

you don't need to replace your hips and knees or feel geriatric trying to walk up the stairs in your seventies.

To understand the importance of release, it's crucial to know what creates restriction in the first place. It all begins with your *fascia*, a system of the body that acts like a densely woven and tight biological sweater. Fascia is a living, extracellular matrix of fibers and water surrounding all of your cells. Your fascia exists as a continuous, gliding structure from head to toe, covering everything inside your skin from your muscles and nerves to your internal organs (such as the heart, lungs, gut, brain, and spinal cord). Fascia connects muscles in chains so they can move together in an organized unit. In fact, I think of the body not as having six hundred different muscles, but as having one muscle divided into six hundred fascial units.

When healthy, fascia is relaxed and wavy, like a sweater with ruffles. It's pliable and can stretch and move as needed. Ideally, your fascia is supposed to glide over your other tissues like a layer of silk. Nevertheless, muscle, fascia, tendons, ligaments, joints, nerves, and organs can get stuck and compromised with fine or thick adhesions, forming scar tissue, knotting up muscles, limiting range of motion, reducing blood flow, and causing inflammation and pain. And it's all connected through your fascia. So tension in your jaw may affect other parts of your body, such as your psoas down near your hips.

When trauma occurs, such as in a car accident, the fascia loses elasticity, may become tight and restricted, and can be a source of tension to the rest of the body. When muscles are involved, such as neck and upper back muscles in a whiplash injury, it can cause pain, reduced range of motion in joints and soft tissues, and poor function. Put simply, muscles become short and stiff, then dysfunctional. It's like a piece of tape was applied to the sweater so that it can no longer bend and stretch when needed. The pain attributed to muscle and the surrounding fascia is called myofascial pain.

When the body's internal connective tissue gets stuck down or, worse, begins to fail, the aging process accelerates and you start to see the telltale signs of it: wrinkly skin, poor vision, bad muscle coordination, and falls and broken bones, among other problems. Perhaps you've already found that tension from dysfunctional movement has brought you to a halt through injury, maybe due to a torn rotator cuff or hamstring. Immobility leads to loss of muscle mass and further aging, not to mention inconvenience and pain. As a result, most of us are dealing with at least one physical limitation—your back goes out, your neck won't turn, you need shoulder surgery.

You could have tension in your diaphragm, which we want to be supple and able to load efficiently in order to help you breathe. The psoas—or abdominal muscles, such as the rectus, transverse, and obliques—may be locked down. If the diaphragm or psoas is stuck, the trunk is less organized as a neuromuscular unit. In general, when your spine, joints, muscles, and tissues are tethered in places, your body may try to compensate, potentially leading to loss of function and unstable positions. Recovery may take longer than it used to. As a result, you may feel tired during or after a workout because of overreaching and underperforming.

Reasons for Chronic Holding and Tension Patterns

Let's start with the main reasons for muscle and joint tension, then jump to the practical and easy ways to release them—actions you can do yourself, anytime, anywhere.

Trauma, surgery, and inflammation are common causes of mild to moderate myofascial restrictions that are unlikely to appear on standard tests, such as X-rays or CT scans. Other times, the tightness is from overuse, perhaps from a pattern of certain muscles taking over for other muscles or nerves to compensate for injury, weakness, or vulnerability. Occasionally, it's due to the body guarding itself in response to a previous vulnerability, such as a person who's had to tackle three-hundred-pound football players (who then fell on top of him), as my husband did in high school. Your body will hang on to the muscle memory, reliving the moment of injury, until you have the awareness and tools to break the cycle. Fortuitously, applying gentle and sustained pressure to myofascial restrictions can allow the fascia and muscles to elongate, sometimes in as little as a single session.

I have myofascial restriction and pain from overwork in my barre and TRX classes, probably related to excess lactic acid and poor alignment, particularly when I'm fatigued in the second half of class and my form suffers. I experience muscular gripping in the right side of my neck from two years ago, when I fainted and hit my head. Some of my muscle spasms relate to old tears; some relate to not getting enough oxygen because I don't breathe deeply enough owing to restrictions in my rib cage. Daily stress contracts my upper trapezius and tightens my neck muscles, even though I no longer sit at a desk to write.

We all have areas of imbalanced musculature, accumulated small tears, habitual tightness, microtrauma, and, in due time, dysfunctional patterns in the body. If ignored, they can snowball into other problems. So if you are not finding relief with releasing your tight places, you may want to review with your physician the common and more rare medical problems that can cause lasting muscle and joint aches:

- Sprains, strains, impingement, other injuries

- Electrolyte issues

- Low vitamin D

- Infections (flu, mononucleosis, Lyme)

- Mitochondrial dysfunction

- Fibromyalgia

- Rhabdomyolysis

- Muscular dystrophies (a group of thirty inherited diseases that damage and weaken your muscles, leading to loss of muscle mass)

- Dermatomyositis (a disease of the connective tissues)

- Polymyalgia rheumatica (inflammatory disorder in older adults)

- Hypothermia or hyperthermia

(Note: It's important that you and your health-care professional rule out more serious causes of myofascial pain before going whole hog into release work. There's a chance that you will not get the full benefits until you fix an underlying biochemical abnormality—that's the deeper, more holistic approach of functional medicine.)

Renske's Cautionary Tale

"I'm the same age as my father when he died," explained Renske in a hushed voice, vulnerable yet articulate, striking a chord in my heart. We were flying to Los Angeles for the weekend. Her comment made me think about how we all want to do better than our parents, at least when it comes to health. You have the benefit of knowing what aging looks like when you witness your parents over the arc of their lives since your childhood, and that knowledge can help you plan accordingly.

Renske's father died, obese and riddled with cardiovascular disease, on the operating table at age forty-two while undergoing his third coronary bypass. She was determined to experience a different fate. She was super-careful about food, exercise, and mind-set. Yet sometimes she overexercised because, like me, she's highly ambitious. Case in point: I was her porter for the weekend because, six months earlier, she'd required surgery for a rotator cuff tear. Unlike my husband, who tore his rotator cuff in a daring snowboarding accident, Renske slowly damaged hers by completing a full Ironman and attending boot camp five days a week. Her damage was cumulative; over months she gradually found that her shoulder no longer performed properly. Her orthopedic surgeon explained that she has the type of shoulder that is more vulnerable to impingement and injury.

In her family, there was a history of poverty, abuse, divorce, fear, paranoia, and disease. Her mother was imprisoned in a Japanese concentration camp for several years during World War II. There was also strength, bravery, and courage, which Renske received in spades, especially evident as she was recovering.

I asked Renske about what had changed in her approach to fitness since her shoulder surgery. Her reply: "I've realized how the care I now give myself and ask for *after* the surgery is probably what I needed *before* the SURGERY." She went on: "I think I 'shouldered' a lot of burdens, but I've realized I can let many of them go."

I first met Renske in a prenatal yoga class when she was pregnant with her first daughter and I was pregnant with my second. Renske now runs a nonprofit food accelerator in Northern California. It's a start-up, so she was constantly attached to her phone throughout our weekend, answering work and family questions.

Renske had to reckon with limitation for the first time in her life. She had more time to *be* while *doing* less. It suits her. She used to think she needed to go to spin class or boot camp every day to stay strong and fit. Since her injury, she's reconsidered what strength is: "Where does strength come from, if not in pure physical form," she asked rhetorically. "If I lose some of my physical strength, do I still have the same internal strength?" We agreed that the answer was yes.

Sometimes the body sends subtle messages. If you're too busy to hear them, they get louder. Maybe it's not until your arm is dangling uselessly and you're unable to lift even a cup of tea that you hear the cry of your dysfunctional shoulder. Sometimes it's utter defeat that finally gets your attention. Don't wait for the cry to get louder; listen in to the quiet calls for help and prevent the more desperate memos from your body parts.

Yoga for Release

Yoga is my favorite way to release my fascia and muscular tissues, but not just any yoga class will do it for me. I practice and teach a form of yoga developed by Ana Tiger Forrest, author of a must-read memoir called *Fierce Medicine*.[4] It's aptly called Forrest Yoga, and teachers can be found globally.[5] Ana Forrest is a maverick sixty-year-old international yoga teacher who has a fresh take on aging: "There's a profoundly negative, degrading attitude in our culture toward aging. I am offering you instead the tools to create a new paradigm whereby, as we age, we can embody and model the beauty of a rich spirit versus just a wrinkle-free face."

When I asked her what was so bad about chronic tight spots, she replied with her characteristic directness: "When energy clogs, matter clogs. It shows up as pain, disease, depression. Do yoga. Feel your emotions. Eat high-vitality food. The clogs clear. The brain smog clears. Life becomes worth exploring." Ana comes from a long line of people with genetic illnesses: obesity, insanity, suicide, cancer, heart attacks. She also had a tough upbringing, to put it mildly. She releases chronic tight spots so that spirit can infuse every cell of her body, because in her philosophy, tight places block spirit from coming in. (I know it sounds abstract, but there

Meet Maureen

Recently I heard a story of release involving a woman with two young children. Maureen was bedridden with Ménière's disease, a condition of the inner ear characterized by vertigo, ringing in the ears (tinnitus), a feeling of fullness or pressure in the ear, and fluctuating hearing loss. Her first symptoms occurred at age twenty-five.

When she first showed symptoms, her conventional doctor put her on antibiotics, then diagnosed her with benign paroxysmal positional vertigo, a condition similar to Ménière's and for which there is no cure. Eventually, she was correctly diagnosed with Ménière's, but it didn't take her long to discover that conventional Western medicine didn't have a lot to offer in terms of solutions. Her Ménière's attacks persisted and worsened to the point that, at age forty-four, she was having two debilitating attacks per month. She tried a lot of modalities to treat her Ménière's, but it was only craniosacral therapy—a precise but gentle touch therapy that releases restrictions all over the body—combined with a low-inflammatory diet (no wheat, low grains, moderate protein, moderate fat, minimal sugar, and lots of greens and other vegetables), exercises that stimulated her lymphatic flow (such as rebounding on a mini-trampoline and hiking), and a few supplements that cured her chronic symptoms.

At Maureen's initial craniosacral evaluation, her therapist observed a seems to be something to it. You can substitute *energy*, *vitality*, or *life force* for *spirit* if that makes more sense to you.)

Even for the most minor tightness, release is essential to prevent unnecessary aging and tension. Release allows your body to enter a restorative phase where it functions more efficiently, effectively, and optimally. That allows your circulatory system, both blood and lymph, to provide fresh nutrients and to whisk away toxins and other biochemical by-products of your internal exposome. You can reestablish correct movement patterns and free up stuck energy to enhance your performance and life force,

complex pattern of tension anchored in the connective tissue at Maureen's pelvic floor. Decades of fascial holding from injury, chronic inflammation, poor posture, and stress caused her pelvis to pull on her respiratory diaphragm all the way up to the floor of her mouth and into her jaw, perhaps contributing to fluid stagnation in the inner ear and inhibiting the natural movement of her neck and skull. Releasing the tension at the pelvic floor was the first priority, even though that was quite distant from where she experienced symptoms. Once her pelvis was more balanced and mobile, Maureen's body began releasing the long-standing patterns that caused vertigo and hearing loss. Over several months, as her body gradually released the chronic tightness, she had more resources to heal herself, and she had more energy for self-care.

Maureen told me: "Creating and following a healing plan that was customized to my needs has allowed me to not only alleviate my Ménière's symptoms and tinnitus, but to feel better at age fifty-one than I have ever felt in my life. Even after putting everything together, it still took my body a long time to undo the damage of the past and relearn healthy functioning. I am still healing and improving all the time. I really didn't know my body could work this well. I am truly in awe of how the body works and that it is capable of so much more than we ever give it credit for." Maureen found a new, better normal.

maintaining muscle mass so that you can slow down aging, and keeping you active and mobile until your last breath.

I've found tremendous relief in opening my own tight places. I no longer have temporomandibular joint (TMJ) pain or spasms in my low

Care of Your Hip Flexors

In the previous chapter, I promised a basic move that will get your hip flexors to work properly again and help keep your belly tucked into your abdominal corset. Your hip flexors are a group of muscles—sartorius, tensor fascia lata, rectus femoris, pectineus, adductor brevis, psoas, iliacus—that contract in order to pull the thigh and torso together, as in a sit-up. Hip flexors may become shortened and tightened if you work them repeatedly and unskillfully, such as by doing abdominal exercises, bicycling, squats, or simply by sitting too much.

The biggest and strongest hip flexor is your psoas major, which connects your low back vertebrae and femur. Your psoas is mighty and deep; some call it the "muscle of the soul." It can be as thick as your wrist in caliber. If the psoas is tight, you may have lordosis, which is a swayback in the lower (lumbar) spine—this problem commonly contributes to low back stiffness and pain, as well as arthritis in the lumbar facet joints. When the psoas is weak, it can cause misalignment that results in tight hamstrings and a more flat lumbar spine and vertical sacrum (instead of a neutral lower spine and sacrum that has a slight forward tilt). When you lose the normal curve of a neutral spine, the low back is weaker when you apply a load, so you're more vulnerable to injury, particularly at the intervertebral discs.

The psoas has a synergistic muscle, called the iliacus, that joins the inner bowl of the pelvis to the femur. The psoas and iliacus work so closely that they are sometimes referred to as the iliopsoas. Collectively, these muscles provide flexion of the hip, as well as its internal and external rotation.

The goal with your hip flexors is to gradually warm them, then strengthen, stretch, and lengthen them. One of my favorite yoga poses to release hip flexors is a combination of butterfly pose into bridge pose (in Sanskrit, supine *baddha konasana* lifting up into *setu bandha sarvangasana*).

back. I've learned how to work different body parts to release my built up and contained energy, so that I have full access to it again. Now I want that for you. Often stretching alone, especially the short time most people devote to it, isn't sufficient to release chronic myofascial tightness.

- Lie on your back with soles of the feet together and knees bent outward and resting on the floor or supported with pillows.

- Tuck your tailbone, and lift your pelvis off the floor. This provides a posterior tilt to the pelvis, which can help lengthen your hip flexors and decompress the low back.

- Roll up as high as you can—it may be one inch or it may be ten inches. Continue to lift the pubic bone up toward the belly button. Slowly wind down and back up five times, inhaling up, exhaling down. This will gradually warm them, sending the signal that you're paying attention and bringing energy and intention to the area.

- Now raise your pelvis into the air and hold for five rounds of breath. As your hip flexors slowly lengthen and release, you may be able to lift higher and go deeper, eventually bringing low back, middle back, and upper back off the floor.

- Repeat three to five times.

Here is a short list of treatments that you can perform on yourself or obtain from qualified professionals in order to release your tight spots and regain full function:

- Stretching—dynamic before workouts, static at other times during the day

- Pilates

- Yoga

- Self-myofascial release with a foam roller or balls (tennis or lacrosse balls or other grippy balls mentioned in Resources)

- Oprah's new favorite: resistance flexibility training,[6] which you can practice on yourself

- Cryotherapy

- Acupuncture (focus on energy flow through meridians in fascia)

- Massage (focus on muscular alignment)

- Craniosacral therapy (CST)

- Skeletal alignment (traction, chiropractic)

- Trauma release (tapping, or Emotional Freedom Technique; Eye Movement Desensitization and Reprocessing [EMDR]; tension and trauma releasing exercises [TRE])

- Other (active release therapy; Feldenkrais; Anat Baniel method; Yamuna body rolling)

Science of Week 4: Release

There is both an art and science to release, but, sadly, rigorous supportive evidence is lacking. That doesn't mean release isn't worth your time. Every day in my work, I see the problem of chronic muscle tension and how it accelerates aging. I notice a lot of patients and especially my husband start off with overuse of a muscle group, such as the upper trapezius, which eventually leads to poor posture, deconditioning of other muscles, and loss of mobility and function. I've experienced the power of release so profoundly in yoga with certain *asanas*, *bandhas*, and *kriyas*, but I don't see the benefits well documented in the medical literature. Release as an outcome is hard to measure objectively in a laboratory setting. So I have to rely more on empirical knowledge and experience to encourage you to release habitual tightness so that you can operate in a more functional state. For the science section here, given the lack of hard data, we will discuss approaches that I've observed to have the greatest impact on healthspan. Even so, I understand there's a leap of faith required of the unconverted. My friend Nick, age thirty-eight, is a great example.

Nick Polizzi, a talented filmmaker and director/producer of the film *The Sacred Science*, suffered miserably with frequent migraines through his midtwenties. With each migraine, he would be out of commission for twenty-four hours. He got sick and tired of strong prescription medications that affected his mood and only worked half the time.

One day, during a migraine, he got a call from his friend Nick Ortner. He had just learned about tapping, or Emotional Freedom Technique (EFT), and wouldn't let Nick off the phone when he complained that it hurt too much to talk. Ortner taught tapping to him right then and there, over the phone. When Nick traced the pain to its root, he uncovered a trauma from childhood that he had forgotten. He experienced a large emotional release, and the pain completely stopped *in a single session*.

Not everyone has such a dramatic release with tapping, but I've personally found it to be helpful when I'm stuck on an emotional or physical problem. It's easy and safe. EFT is proven to help relieve pain from tension headaches.[7] Check out other science-based benefits of tapping:

- EFT lowers anxiety, depression, and cortisol, some of the biochemical drivers of tension in the body.[8]

- In women with fibromyalgia, EFT reduces pain and anxiety and increases activity.[9]

- In a systematic review, EFT was found to improve post-traumatic stress disorder (PTSD), phobias, test anxiety, and athletic performance.[10]

- EFT was shown to be superior to diaphragmatic breathing, progressive muscular relaxation, an inspirational lecture, and a support group.[11]

John Upledger, DO, who developed CST, had a great sense about why release is important: "The secret something that is shared by all effective healing methods is the process of leading the patient to an honest and truthful self-discovery. This self-discovery is required for the initiation and continuation of self-healing. It is only through self-healing—in contrast to curing—that patients can experience both permanent recovery and spiritual growth."[12] While Dr. Upledger is referring to something bigger than a technique, the takeaway is that the best healing occurs when you meet the modality in the middle and activate your own healing powers through releasing myofascial patterns that no longer serve you, actively creating a process of rest and renewal. It's about connecting to what Upledger calls your inner physician.

Initially I was skeptical, but what I've seen and witnessed in my own body make me a believer.

Practical Myofascial Self-Treatments

Stretching

The best thing you can do this week is start making release techniques a daily ritual. Most people agree that stretching is a good idea—after all, it promotes flexibility and range of motion and lessens your risk of strain. But that's where the consensus ends. It's not clear how much you should stretch, how long to hold each stretch, and how many times per week you should do it. Sadly, stretching has been studied less thoroughly than other types of exercise and movement, so evidence is limited. Despite that, we can draw a few conclusions so that you can enhance your balance, prevent falls, and even relieve arthritis and pain in your back, knees, and hips.

- Healthy adults should perform flexibility exercises such as stretching, yoga, or tai chi at least two to three times per week for all major muscle groups, including neck, shoulders, chest, trunk, lower back, hips, legs, and ankles.

- Ideally, spend sixty seconds on each stretch for a targeted muscle/tendon group. Stretching should create a gentle tug but no sharp or radiating pain.

- Stretch throughout the day but not before a workout—yes, the rules have changed. Static stretching, such as touching your toes, is no longer recommended *before* a workout because it can cause injury, doesn't prevent muscle soreness,[13] and may inhibit maximum muscle performance.[14] Instead, experts recommend *dynamic stretching before a workout*—where you move in a way that lengthens muscles and connective tissue—such as a set of ten or twenty-five jumping jacks.

- Make sure to stretch *after* a workout, as described above.

Self-Myofascial Release

Here's what we know about self-myofascial release. In the short term, self-myofascial release, which you can do with a foam roller or tennis ball, boosts flexibility and reduces muscle soreness without harming athletic performance. Self-myofascial release helps the range of motion of your joints and is best performed after a workout.[15] Self-myofascial release may improve function of arteries and endothelial cells (the cells that line blood vessels of all types), and it may optimize parasympathetic nervous system function, which could aid recovery. Then again, it's not clear that self-myofascial release enhances flexibility long term.[16] In people with low back pain, self-myofascial release of the transverse abdominus, one of your deep belly muscles, improves the integrity of the musculofascial corset system of the belly, which is important for spinal stability.[17]

I'm a fan of self-myofascial release for the diaphragm. You may know that the diaphragm attaches on the lower six ribs, like a skirt steak. The tail of the diaphragm crosses the psoas to attach to the lumbar spine; that's why shortening and stiffness in the diaphragm can lead to decreased mobility of the ribs (and feeling like it's hard to take a full-lung-capacity breath), low back, and hip pain.

Jill Miller of Yoga Tune Up believes that restrictions in the diaphragm can make it hard to calm down the nervous system. She advises using small grippy balls (see Resources) or two tennis balls and placing them under the midback. Lie over the balls on your back, with the balls parallel and just to the left of your spine. Roll up and down and back and forth until you feel a sensation of release. Then repeat on the other side. When performing self-myofascial release, avoid rolling over bone or swollen tissues and stop if you feel any sharp, shooting nerve sensation.

Myofascial Release Performed by Trained Professionals

Other forms of myofascial release may be performed on your body by a practitioner. It's not a one-size-fits-all approach; you have to experiment to find the right fit for you. Often the best choice is a form of release that matches the specific form of movement and exercise you're getting. Some types are less known by the general public, including cryotherapy, Rolfing, chiropractic, and CST, which I'll describe next in greater detail. From a scientific perspective, some of the time-honored traditions for recovery and release include cooling (icing or cold therapy), massage, and compression to hasten the regeneration of the neuromuscular junction. Among these, cryotherapy may prevent or minimize muscle soreness after a workout,[18] presumably by restricting blood flow to the affected muscles and reducing inflammation so that repair may occur.[19]

Craniosacral Therapy

I'm trained to be a skeptic, but the most effective treatment I've found for the problem of being stuck in your tissues is CST, an alternative treatment that releases restrictions in the fascia and fluid around the spinal cord, cranium, and throughout the body—and subsequently restores body function.

Two years ago, I fainted in a standing position and hit the back of my head and neck on a stovetop. For weeks afterward, I was foggy, with a stiff neck and weird twitch of my head to the right. A massage therapist recommended that I receive CST, and a friend referred me to Robyn Scherr, a gifted craniosacral therapist in Lafayette, California. I had my doubts about CST and knew the scientific proof was limited, at least in the eyes of standard medicine.

In my first session, Robyn started palpating the left side of my neck. I had explained that I experienced chronic pain on the right side of my neck, so I thought perhaps she hadn't heard me. But as she palpated deeper into my tight left neck, I felt a very clear sensation of fluid releasing in my left neck, like a water balloon had burst.

She saw my eyes fly open and asked, "Did you just notice something?"

"Yes, what the heck was that?"

"You just released an energy cyst, a localized area of compressed energy," Robyn explained. "When energy enters the body in overwhelming quantities (or with an overwhelming quality), the body adapts to the presence of this energy by trying to contain it. It compresses the foreign, disorganized energy into a small space, thereby creating a cyst of energy. It's the body's way of minimizing disruption. Your body works around an energy cyst until it has the resources to deal with and release the effects of that injury.

"An example of an overwhelming quantity would be the force that entered your head and neck when you fainted and fell. It was definitely more than your body could handle at the time. An overwhelming quality, for example, could be an emotion. That's why minor physical injuries, when the emotions surrounding them are intense, can tend to linger."

Uddiyana Nurtures Release

More than twenty-five hundred years ago, the Yoga Sutras explained that yoga is the key to longevity, especially with expert use of the *band-has* (energetic locks) of the body. The idea is that mastering the locking and unlocking of the *bandhas* can slow down aging. Practicing *bandhas* is one of the best ways to release chronic holding, myofascial tension, and even psychological trauma. My favorite is the abdominal lock called *uddiyana bandha*; *uddiyana* means "to fly up or soar."

The practice is to pull the abdominal muscles upward toward the spine after exhalation, and to hold the exhale as long as possible. Avoid in pregnancy or if you have any of the following: high blood pressure, heart disease, hernia, glaucoma, gastrointestinal ulcers.

Here's a primer on *uddiyana bandha*, performed first in a seated posture, then in bridge pose.

1. Get a feel for *uddiyana.* Sit comfortably with knees bent. Inhale and exhale deeply. Round your torso forward. Pause at the end of exhale, lips closed; tuck chin toward chest, and contract abdominal muscles upwards toward thoracic spine. Hold as long as comfortable, from ten seconds to one minute. Then release chin lock and abdominal muscles, and gently inhale. Repeat a few more times.

2. Try *uddiyana* in bridge pose. Lie on your back with knees bent and feet flat on the floor. Inhale and lift tailbone, low back, midback, and upper back into bridge pose. Spread the ribs and take a full inhale and exhale. Pause and hold after the exhale, lips closed and chin tucked toward chest. Lift your belly wall and organs toward your midback. It should feel like you're creating a vacuum in your belly. Hold the exhale for as long as you can, from ten seconds to one minute. Gently release chin lock and abdominal muscles, and inhale.

3. Soften belly, and repeat *uddiyana* two to four more times.

By performing *uddiyana*, you will wake up dormant tissues that are underused, such as the innermost intercostal muscles and the deep abdominal muscle layers. Once you know how to practice *uddiyana*, you can perform it seated in cross-legged position, in dolphin pose, or in almost any other *asana* of your choice where freedom of the diaphragm would be helpful.

It all sounded a little strange to me, but I couldn't get over the fact that I actually felt a cyst release and I could move my neck more to the right and left. My neck was soft, and it didn't hurt anymore. My favorite part about CST is that the positive shifts tend to stay put. I'm now fully recovered from that head injury. I continue to receive treatment about once every month or two, more often if I have areas of tightness, such as when I've been traveling a lot.

The Importance of Feeling Safe (When You Are Safe)

The tendency to hold tension in the body, known as being in sympathetic dominance or in the fight-or-flight state, is meant to keep us aware and alert. The fight-or-flight system saves our lives when we're in danger—for example, by moving us out of the path of a speeding car—and propels us to greater performance when we're under work deadlines or during rigorous workouts. This response system is useful; we need it. Sad to say, the nervous system often doesn't calm down (downregulate) when the original danger has passed. If it did automatically relax when a threat was over, we wouldn't hold tension! In our society, we are out of balance; the majority of us spend most of our time in sympathetic dominance, which ages the body, and not enough time in autonomic balance, whereby the rest-and-digest, tend-and-befriend mode restores the body.

So it's crucial to pay attention to and tend to our relaxation systems to counterbalance fight-or-flight. To turn on parasympathetic tone, you need safe places to relax and, when you're with a body therapist, safe hands. Then the yoga can work, the self-myofascial release can work, and the aromatherapy can work. As Bessel van der Kolk says in his excellent work *The Body Keeps the Score*: "Interventions are successful if they draw on our natural wellsprings of cooperation and on our inborn responses to safety, reciprocity, and imagination."[20]

Once we can recognize that we are not in danger, not under pressure, and don't need to rush, we are better able to move out of fight-or-flight and into rest-and-digest. Our bodies can then make the materials that keep

Meet Mary, a Fifty-Eight-Year-Old Newly Minted Yoga Teacher

Mary came to see me six years ago for high stress, a racing mind, and insomnia. That's when I added to her regimen oral progesterone and Cortisol Manager, both of which help clear tension in the body so that you can sleep. Mary was a nervous person who worried about her kids, finances, and what she was going to wear to the local school parties. She worked full-time in financial planning, selling securities, until the hours became too demanding to keep up with her children's needs. She took her first yoga class at age forty-eight when she cashed in a Groupon with a friend. She had never been an athlete, so she had the stamina for only two classes per week, but her first month of yoga spoke to her in a way she had never experienced.

"At the time I had three kids at home and was in a very difficult, unfulfilling marriage. Finding a few hours a week on my yoga mat brought me beautiful exhilarating peace. I loved the connection of mind, body, and spirit. Within six months I was attending four or five classes a week, and within a year, seven classes a week. I planned my days, weeks, and vacations around where and when I could attend classes. Within four years, at the age of fifty-two, I became certified at the two-hundred-hour level to teach with YogaWorks."

Mary went on to secure additional yoga certifications at the five-hundred-hour level. She has been teaching and practicing yoga consistently ever since. Yoga has been a major source of release for her: "Physically, yoga has had the most effect on my neck and hips. I have been in three car accidents over the past twenty years, all hit from behind with resultant whiplash. Yoga keeps my range of motion intact and helps me relax and release this area."

Mary has a healthspan score of 85.

us more supple, repair injury, and promote flexibility in our muscles and elasticity in our tissues.

Protocol for Week 4: Release

When the body's internal connective tissue gets stuck down or, worse, begins to fail, aging accelerates. Besides hindering your mobility and adding to your stress, chronic tightness leads to more wrinkles, uncoordinated muscles, and declining vision. So even if you already have your ways of releasing restricted pathways in your body, pick one new strategy to try this week. You might try dynamic stretching before a workout, cold-water immersion, or a visit to a craniosacral therapist.

Basic Rituals

- Stretch (static or dynamic) daily for at least ten minutes. Work all of the major muscle/tendon groups including neck, shoulders, chest, trunk, lower back, hips, legs, and ankles. Spend sixty seconds on each stretch. If you are at a loss, start with the side bend yoga pose described here:

  - Begin comfortably seated in a cross-legged position, feet active (feet flexed under opposite knees). Close your eyes if you can, soften your breath, and feel your inside.

  - Inhale and lengthen your spine toward the sky; exhale and raise left arm above head, hand active, and side bend to the right as you walk your right hand a few inches on the floor from your right hip.

  - Turn your left hand so the palm is facing you and open your hand bones by extending through your fingertips.

  - Release both shoulder blades down your back.

  - Inhale and exhale for a total of seven rounds of breath.

  - Inhale back to your midline, and repeat on the left side.

- Perform self-myofascial release, such as savasana (corpse pose) over a tennis ball or rolling out a sore area with a foam roller. My husband has a date with the foam roller every night at about nine thirty, before we climb into bed.

Supplements

Many people are deficient in relaxation minerals such as magnesium.

- Magnesium counters the stress response, helps your muscles release, and may even enhance your sleep. Sometimes your tightness or stiffness is a sign of magnesium deficiency. It's needed for hundreds of biochemical reactions in the body. Foods that are rich in magnesium include kelp, dulse, almonds, cashews, Brazil nuts, pecans, walnuts, collard greens, shrimp, avocados, and beans. Take 300 to 1,000 mg per day, unless you have kidney disease, in which case you should consult your healthcare practitioner.

Advanced Projects

- Traction. I perform traction several times per week by hanging from a pull-up bar for sixty seconds. Barre classes often have a bar that you can use after class, or you can buy an inexpensive pull-up bar online. Your hang will start to lengthen the muscles of your low back, realign your spine, and strengthen your handgrip, an important marker of aging. Start with a thirty-second hang. As you get stronger, aim for sixty seconds or longer. Skip the hang if you have a shoulder injury or are pregnant.

- Cryotherapy. There are many ways to cool down your body as a form of hormesis (the beneficial biological response from exposure to a harmful agent in low doses), in this case ice or cold temperatures, which could be lethal at high doses (causing frostbite and potentially death).

  - Self-experimenter Tim Ferriss performs ten-minutes ice baths, described in detail on his blog, to induce sleep.[21] He describes them as "getting hit with an elephant tranquilizer." Here's how: Buy two or three bags of ice from the grocery store and put into a bath until 80 percent melted. Start by immersing the lower body only, then spend the second five minutes with your upper torso submerged in the ice bath. Apparently, it may boost fat loss, but I haven't seen great results in women.

  - Other people swear by a new fashion trend: wear an ice vest. To quote *The Atlantic:* "When you first put on the ice vest, you will feel cold. Not intolerably cold, but cold enough to make you think, *What am I doing with my life?*"[22]

  - Or you could outsource the low temperature. My friend and colleague Alan Christianson, N.D., swears by a cold sauna for lowering inflammation. I've tried it a few times and found it interesting, but as you'll learn in chapter 10, I have a gene that makes me wig out at cold temperatures, so I'm not the best study subject.

- Emotional Freedom Technique. You can immediately start using EFT to tap on acupressure points in your body and release tension or trauma. As you tap, you'll voice positive reminders while you activate the same energetic meridians used in Chinese medicine and acupuncture but without the needles. This is a great tool to have in your tool kit for on-the-fly release.

– How to Tap—the Mini-Version

- Prepare. Remove glasses and watch. (They may interfere with the subtle energetic shifts you want to create.) Trim your nails so your fingertips can make contact with your tapping points without hurting.

- Rate the intensity of your target issue from a scale of 0 to 10 (10 is the most distress you can imagine; 0 is none). You'll be venting about how you feel as you tap, voicing a short phrase that's relevant to current issue you're feeling in your body that you want to release, such as "My hip hurts" or "Parenting has been very stressful lately."

- Create an affirmation, such as "Even though I feel overwhelmed and stressed, I accept myself," or "Even though my neck is tight and stiff, I choose to feel calm and relaxed."

- Review the three tapping points in the following sequence:

  – karate chop (side of hand)

  – under eye

  – collarbone, just below the hard bone

- Start with the side of the right or left hand, called the *karate chop* point (the fleshy part), and tap rapidly with your four fingertips of the other hand. Use your four fingers as a group. Aim to tap five to seven times, voicing out loud your affirmation three times.

- Next tap under your eye, voicing how you feel in a short phrase noticing what you're feeling that you want to release.

Lie on Balls for Five Minutes

One of the fastest ways to age your body is to engage in time compression, the feeling you get when you don't think you have the time to accomplish all of your goals in a day. It can ramp up your fight-or-flight response well beyond what's appropriate for the situation. But five to ten minutes of a restorative pose can counterbalance a high-pressured day. I like to do restorative poses over balls that are placed along the spine to release tight spots. Rolling on two balls placed under the upper and midback may free up your diaphragm and increase the capacity of your breath volume, which will oxygenate the blood and connective tissues and activate the vagus nerve, creating a sense of calm and release. The vagus nerve is the portal to the parasympathetic nervous system that governs restorative bodily functions and where most healing and release occurs. I learned this technique from Jill Miller and one of her certified trainees.

The goal is to quiet the mind and unlock tension in the diaphragm and low back. Here's how to do it.

- Lie on your back with your knees bent.

- Place two balls (tennis or lacrosse balls) under the lower back, about halfway up your spine, one ball on each side of the spine.

- Inhale and tuck pelvis up. Exhale, and drop tailbone over balls.

- Inhale pelvis up; exhale pelvis down.

- Move your balls up about one or two inches toward the midback. Shift hips toward the right, and drop the right buttock toward the floor. Then shift hips to the left over the balls, and drop the left buttock.

- Move the balls up higher, to the mid-to upper back. Straighten out your legs and extend your arms, palms facing up. Breathe slowly and deeply for five breaths.

Tap just one side or both sides at once (you'll activate the same meridians). Move next to the collarbone. Repeat the sequence three times.

- Tap back where you started, to complete the sequence.

- Cannabidiol (CBD) oil. No, I don't want you to become a pothead. CBD oil is the nonpsychoactive part of the Cannabis sativa plant. It's been used for centuries for conditions such as gout, rheumatism, pain, anxiety, and fever, and it's now under investigation as a neuroprotective, antiepileptic, antispasmotic, and anti-inflammatory agent.[24] It is available in the United States as an over-the-counter treatment. I recommend starting with doses of 5 mg up to three times per day like Rosalie in chapter 6.

- Outsource your release. Consider visiting a practitioner who can perform massage therapy, Anat Baniel, Rolfing, TRE, resistance flexibility training, CST, or acupuncture.

Your Daily Routine

The next page shows what a daily routine using the basic rituals from weeks 1–4 might look like; it's based on how Renske structured her weekdays after her shoulder surgery. Make it your own!

Recap: Benefits of Week 4

As my craniosacral therapist, Robyn Scherr, eloquently states: "Our bodies work around held experiences until the load becomes too great; that's when we develop symptoms." Whether your load is too great, and symptoms have started, or your load is still manageable, the short-term benefits of release are increased flexibility and mobility. Release allows you to age more gracefully while reducing your physical stress. As I described in the introduction and chapter 7, aging begins in the muscles with loss of mass and more limited range of motion, and release allows your muscles to fire

A TYPICAL DAY IN THE YOUNGER PROTOCOL: RENSKE

6:30 A.M.	Wake up and apply skin care
6:45	Drink coffee, check e-mail/text, eat breakfast
7:00	Review plan for the day, meal plans, etc.
	Take supplements (multivitamin, vitamin D, probiotic, omega-3) and commit to drinking more water today
7:45	Work
9:00	Exercise 60 minutes, starting with shoulder release and strengthening, slowly adding yoga back in
10–4	Work
	Lunch 30 minutes
	Siesta/quiet time 20 minutes
4:00 P.M.	"Domestic blocking and tackling" as Renske puts it—managing the amount of work it takes to run a household and raise kids
6:00	Dinner
7:00	Wind down (TV 1x a week, reading, news)
	Release shoulder/perform exercises
9:00	Shut down electronic devices
	Face/teeth routine — cleanser, serum, oil, flossing, brushing
	Take evening sleep tincture (relaxation minerals and herbs, including magnesium, calcium, passionflower, and valerian)
10:00	Lights out

more effectively so that you are strong and powerful and can use your body to its full function.

Bottom Line

There's no downside to releasing chronic tension in your muscles and fascia. This week offers a protocol that you have no excuse to skip; you can perform simple attuning and stretching any time. Find your favorite method and stick to it this week, and keep performing that method two to three times per week for the remainder of the Younger protocol.

CHAPTER 9

Expose
WEEK 5

Let us try to teach generosity and altruism, because we are born selfish. Let us understand what our own selfish genes are up to, because we may then at least have the chance to upset their designs, something that no other species has ever aspired to do.

—**Richard Dawkins**, *The Selfish Gene*

There are toxic chemicals lurking in your home. They're even accumulating inside your cells. Every day, you are exposed to toxins, even if you aren't aware of them. Other toxic exposures, such as severe childhood trauma, can sometimes age you as much as low-level exposure to mold in your home, synthetic skin-care products, or pollution. No one is immune: even newborn babies are bombarded through the placenta with chemical compounds present in air, food, water, soil, dust, and consumer products.

I apply only organic skin-care products and clean my home with green cleaning supplies, so I was astonished when I received my lab results. The lead and mercury levels in my urine and blood were through the roof. I had to dig deep to identify the culprits, which turned out to be the lead in my green tea, tap water, and favorite lipstick and the methylmercury from fish I'd been eating.

You may not notice the exposure, but over time, you can measure the synthetic chemicals in your blood, urine, and hair to determine what has entered your body from the environment. Sometimes the original toxin is the worst part. Other times, it's the toxin's metabolites, substances produced when your liver chemically alters the original toxic compound into even worse chemicals. As you learned in the introduction, genetics accounts for only 10 percent of disease; 90 percent of it is due to environmental exposures. Toxins can damage your cellular DNA, both in the nucleus and in the mitochondria. We're exposed every day, sometimes all day, so there's no luxury to detoxing and decontaminating—it's crucial.

Toxic exposures can accelerate your aging or, once discovered, be grist for the mill of higher aspirations, such as transformation and resilience. You must choose to be proactive. If you don't choose, your body does the best it can when assaulted by toxic chemicals, but the toxins usually win—they harm your mitochondria and get stored in your fat. Generally, your lifetime of exposures ages you prematurely, making you more brittle and brimming with oxidative stress. Oxidative stress refers to the biological imbalance between your body's production of free radicals, which damage DNA, and your body's capacity to neutralize them with antioxidants. If you have too many free radicals and not enough antioxidants (perhaps because you rarely eat broccoli or take vitamin C), then you develop excessive oxidative stress. Alas, most people lack sufficient antioxidants, and free radicals continue to run rampant in the body.

Excessive free radicals and oxidative stress are like the bullets flying in the bad neighborhood for your genes that I described in the introduction (page 3). Think of oxidative stress as drive-by shootings from free-radical damage that cause genetic mutations; heavy-metal exposures as armed robberies gone wrong; and mold in water-damaged buildings as a carjacking at gunpoint. Who would stay in a neighborhood that bad? I know I wouldn't.

I can help you reform the neighborhood no matter how far gone it is. The environment—everything from the air you breathe to the foods you eat to the trauma you've experienced—can either help or harm your body.

It's critical to learn more about the most common bad exposures, why they age us, and what can be done, including the ways to increase your positive exposures. Epigenetically, there's a lot happening in this week of the protocol. Here's the summary of the gene-environment interactions. I know it all sounds complicated and overwhelming—I promise there won't be a test! The takeaway is to know that quite a few of your genes determine how risky your daily exposures are.

- MTHFR is the Methylation gene that helps you produce a usable form of vitamin B_9 as well as detoxify alcohol and other toxins.

- GSTM1, glutathione S-transferase mu 1, codes for an enzyme that makes the most powerful antioxidant in the body, glutathione. There are at least eight forms of glutathione S-transferase genes, but GSTM1 is the polymorphism that I inherited that makes me accumulate mercury. For now, just remember that glutathione is good!

- GPX1 codes for another glutathione enzyme, glutathione peroxidase 1. It's one of the most important antioxidant enzymes in the body and helps to detoxify hydrogen peroxide, a reactive oxygen species.

- SOD2, sometimes MnSOD, for manganese-dependent superoxide dismutase, is the gene that codes for superoxide dismutase 2. It helps to heal the mitochondria from oxidative stress and it prevents senescence, the zombie-like state that can accelerate aging.

- CAT, catalase, a gene that protects you from oxidative damage.

- NQO1, the gene for NAD(P)H dehydrogenase, quinone 1, involves a supplement you may have heard about, coenzyme Q10. It is another important antioxidant that prevents free radicals from exerting damage. This is an important gene for preventing cancer, Alzheimer's disease, and liver damage from toxins such as benzene, a constituent of crude oil that acts as a carcinogen.

- FOXO3, one of the longevity genes described in chapter 1, is involved in ovulation and how fast your eggs ripen; it protects you from oxidative stress and regulates growth factors in your skin.

- MMP1, matrix metalloproteinase 1, regulates calcium signals in the cell and collagen breakdown, so it's very important for keeping your skin young-looking.

- Other genes involved in exposure: CRP, DAO, EPHX, HNMT, Mold (HLA DR), PYCR1, among many others.

Why It Matters

Since mapping the human genome, scientists developed an important complementary concept called the exposome—the sum of all exposures in an individual over a lifetime from diet, lifestyle, and behaviors, how the body responds to them, and, finally, how these exposures relate to health.

Understanding your exposome requires that you be able to measure exposures and their effects on the body. Your genes produce specific biomarkers that can be detected in your blood, urine, and hair. Biomarkers indicate the effect of an exposure, susceptibility factors (including genetic susceptibility), and disease progression or reversal, and they can even help identify the best treatments in some diseases, such as breast cancer. Biomarkers help health professionals accurately measure exposures and their effect, although it's not necessary to perform expensive testing before you start the inexpensive cleanup of your body. You can remove the toxins and exposures that are most likely negatively affecting your body and healthspan, leading to premature aging. Your lifestyle changes can reverse this trend. The protocol is designed to help you reset your exposome by altering the most common exposures, both good and bad.

Take my body as an example. I mentioned before that I was conceived when Twiggy was popular in 1967, and my mother looked a lot like her while pregnant with me. At five foot seven inches, she gained only twenty pounds over the course of her full-term pregnancy. I was six pounds at

Signs of Toxic Exposure

Nearly every part of the body may be affected by chemical toxins, so creating a list of what to watch for is tricky. If you're concerned you've been exposed to a toxin, discuss it with your health-care professional, who can order additional testing.

- Flulike symptoms (fatigue, sore throat, upset stomach, fever, earache, headache)
- Muscle spasm, pain, and cramps
- Joint pain, particularly back, foot, wrist
- Bone pain or density loss
- Gastrointestinal symptoms such as nausea, vomiting, bloating, and/or diarrhea
- Fatigue
- Brain fog
- Visual loss, particularly in peripheral vision or night vision
- Dark circles under the eyes
- Sore throat with or without enlarged lymph nodes
- Dizziness
- Numbness in hands or feet
- Twitching eyelids
- Gingivitis (inflamed gums)
- Changes to nails or skin, or hair loss
- Rashes, hives
- Hormone changes
- Temperature dysregulation
- Cold hands and feet
- Failure to sweat
- Dry eyes or mouth

birth. Paradoxically, low birth weight is associated with a later struggle with excess weight. In total, these factors may have turned *on* my obesity, faster aging, and type 2 diabetes genes.

- I was born vaginally (good for my microbiome) and was breast-fed just two months, which was better than nothing. I grew up eating Pop-Tarts and Ho Hos (not good for my microbiome) until my mother started reading Adelle Davis and following her food rules. Suddenly, my lunch box was stuffed with sandwiches made with dense brown bread, homemade almond butter, and strawberry jam. No one at school wanted to trade lunch items with me anymore.

- Growing up in Annapolis, Maryland, I remember riding my bike in late summer behind trucks that were spraying a chemical to reduce mosquitoes, probably an insecticide like DDT. Most likely,

CELLULAR AGING

EXTERNAL ENVIRONMENT
- Stress
- Poor connection or community
- Toxic home or work (mold)
- Lifestyle
- Infections
- Drugs
- Air pollution

INTERNAL ENVIRONMENT
- Active and stored endocrine disrupters
- Mitochondrial damage
- Inflammation
- Stem cell exhaustion
- Dysbiosis/gut permeability
- Senescence
- Pre-existing disease

BIOMARKERS
- Blood sugar
- ALT (liver enzymes)
- IL-6, hsCRP, homocysteine
- Heavy metals
- Metabolic waste products/hormones
- Immune modulators
- Persistent Organic Pollutants (POPs)
- Telomere length

the mosquito spray was an environmental stressor that may have turned on a heightened risk of breast cancer in my body.

- Fast-forward to years of medical training. I thought I was virtuous to cook meals in advance and cart them to work in plastic containers. Little did I know that the plastic leached chemicals into my food and interfered with my hormonal function, even at extremely low doses. I'm not the only one concerned about fake hormones, called endocrine disrupters; the Endocrine Society issued a new scientific statement about the far-reaching effects of this exposure and what we need to do about it.[1]

- In 2006, I discovered something unsettling about my lunch at work. I was postpartum, having recently given birth to my second daughter, and I had cravings for tuna sashimi. I thought I was being virtuous by skipping the rice, but little did I know that what seemed to be a reasonable consumption of fish was actually making mercury accumulate and cause heavy-metal toxicity inside me. I was tired all the time, but I blamed it on the sleep deprivation of caring for an infant. I felt brain-dead, but that's normal for a tired new mom. I was fat and had trouble losing the baby weight. I had gingivitis (inflammation of my gums), and my dentist said it was probably the recent pregnancy. Then I tested my mercury level and found I was off-the-chart high. I got rid of it as fast as possible with oral chelation, but that meant I had to switch to salmon instead of tuna. I have the genetic variant of GSTM1 that makes me accumulate mercury. Half the population is like me and missing the normal gene for detoxing mercury and other poisons; you may be one of them.

I'm not alone in my exposures, and they extend far beyond childhood. So let's expose these problems and start cleaning up your neighborhood.

You have your own unique exposome, but chances are that a few exposures have sent your genes on a downward spiral in terms of supporting your healthspan. It's not your fault; something went terribly wrong starting in the Industrial Age. After thousands of years of living in harmony with the environment, people started to destroy the aspects of the environment that healed. Big Chemistry started to grow, and new synthetic chemicals were considered innocent until proven guilty.

From 1900 to 2000, the average life expectancy increased by thirty years in the United States and in most developed countries. This showed progress in many ways, but it also meant more years of environmental exposures.[2] Since your lifetime is likely to be quite long, limit the damage. Detoxify old exposures and prevent future ones that may take you down the path of disease.

Science of Week 5: Expose

Let's zero in on the science of exposures that affect aging, inflammation, and degeneration—including exposures that affect your skin, brain, weight, and breasts—and look at what will improve your own prevention and reverse the exposure-disease interplay, which results in disorders such as cancer, diabetes, heart disease, Alzheimer's disease, even autism.[3] In order to understand the causes and learn how to prevent disease, we need to establish the environmental cues that lead to disease, accelerate the aging process, and put you at risk for accumulation of toxins, wrinkled skin, osteoporosis, and mold illness. If you're not interested in the science, skip ahead to the protocol for this week's rituals that will help reverse your negative exposures and add some positive exposures that will turn on the best genes.

Environmental Exposures from Products and Chemicals

Here's a sobering statistic: one in every three people right now will be diagnosed with cancer over his or her lifetime. Every minute, more than fifteen people die of cancer around the world. Breast cancer is the most common type of cancer affecting women worldwide, and many cases are caused by the interaction of your genetics with the environment and, as a result, your exposome. What's even more sobering is that cancer is mostly preventable. Many environmental exposures affect your risk of breast cancer:

- Ionizing radiation such as from X-rays, CT scans, and long flights on planes (from radiation in the atmosphere, although the greatest risk is for pilots and flight attendants)

- Synthetic hormones from taking hormone therapy or hormonal contraceptives

- Certain female reproductive factors, some of which are within your control, some of which are not (early puberty, hormones for infertility, late menopause, never breast-feeding)

- Alcohol and other dietary factors

- Obesity

- Physical inactivity

- Artificial light at night (ALAN)[4]

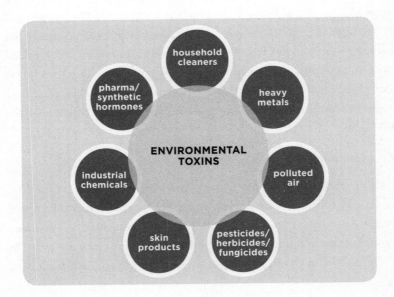

Dr. Sara's Blacklist

By now I am certain you want to know the synthetic chemicals to remove from your life. I made a list for you.[5] These are the most common toxins you'll encounter in your skin products, home, and work environment. While it will be nearly impossible to remove all of these, it is important to be aware of the dangers listed below to become more intelligent consumers for the sake of your health. (See alternatives in Resources.)

1. **Skin products.** No one wants deeper wrinkles, hollows under the eyes, or loss of volume in the cheeks. Most women use nail polish, hair dye, and lipstick. Yet too many cosmetics are created from materials that damage the body. Not only that, these products have been designed to penetrate your skin's innermost layers in a way that encourages toxic ingredients to end up inside your body. Are any of these in your medicine chest?

   - **Lead.** A neurotoxin commonly used in lipstick and in dark hair dyes, lead is a dangerous toxin that damages cognitive ability, and significant exposure can result in stroke and heart disease.

   - **Phthalates.** Phthalates are found in a surprising number of common household products: shampoos, deodorants, body washes, hair gels and sprays, and nail polish, to name just a handful. They cause birth defects in male fetuses and are associated with poor egg quality and early menopause in women. In addition, there are direct links between phthalates and breast cancer and type II diabetes.

   - **Parabens.** Parabens are a group of preservatives found in about 85 percent of cosmetics to prevent the growth of microbes such as yeasts, molds, and bacteria. Most commonly used in deodorants, antiperspirants, shampoo, conditioners, lotions, cleansers, and exfoliators, parabens are so prevalent that a recent survey by the CDC found traces of parabens in all Americans. Parabens are linked to endocrine, reproductive, and developmental problems.

   - **Sodium lauryl sulfate (SLS).** SLS is a toxic detergent commonly used to create lather in shampoos, soaps, and toothpastes. The suds are nice, but not worth the skin irritation, hair loss, and risk of breast cancer and male infertility that come with it.

– **Triphenyl phosphate (TPHP).** TPHP hardens plastic, and it's used to make nail polish resist chipping and as a fire retardant in furniture. Science shows that TPHP affects hormone nuclear receptors,[6] may change the balance of sex hormones,[7] and can be toxic to liver cells.[8] One study found that TPHP (sometimes TPhP or TPP) leaks into the bodies of women who wear nail polish. Look for TPHP-free nail polish.[9]

2. **Kitchen and cleaning supplies.** Here are a few of the chemicals that you'll find in your bathroom, laundry room, and kitchen. They bind to your hormone receptors, bio-accumulate, and cause an assortment of health symptoms.

   – **Alkylphenols.** Used to make detergents, fuels, lubricants, and plastics, and found in tires, adhesives, coatings (such as in canned foods), rubber products, and carbonless copy paper. One type, bisphenol A (BPA), found mostly in plastic containers, is known to disrupt estrogen, thyroid, testosterone, and insulin function. Ditch the plastic water container right away.

   – **Fluoride.** We are familiar with the benefits of this chemical agent in the reduction of tooth decay. Nonetheless, excess fluoride has been proven to weaken bones and negatively impact brain function.

   – **Other toxins in drinking water.** Besides fluoride, you may find in your drinking water lead, pathogenic bacteria, byproducts of chlorination known to cause cancer and reproductive problems, arsenic, and—my favorite—perchlorate from rocket fuel.

   – **Organophosphates.** This pesticide is found in the air, the soil, and the food we eat. One Harvard professor estimates that we've lost 17 million IQ points from organophosphate exposure.

3. **Building materials.** These toxins also lurk where we live and work.

- **Asbestos.** This mineral was used as electrical insulation because of its resistance to fire and heat. Asbestos has been banned and phased out since the 1980s, but it is still found in older homes and buildings. Inhaling asbestos causes serious and sometimes fatal lung problems, lung cancer, and mesothelioma.

- **Cadmium.** An element used as a pigment, an anti-corrosion coating on steel, and a plastic stabilizer. It's listed on the European Restriction of Hazardous Substances, but it's still being used in solar panels, fossil fuels, iron and steel production, cement production, phosphate fertilizer, and foods such as bread and vegetables. Classified as carcinogenic, cadmium is associated with breast, lung, prostate, and kidney cancer. Those at greatest risk for cadmium toxicity are postmenopausal women with low iron.

- **Formaldehyde.** Found in particleboard (often used to make kitchen cabinets) and, believe it or not, Brazilian blowouts. Yes, the particleboard under the kitchen cabinet and your latest hairstyle both share the same toxin that has been linked to cancer of the upper throat and bone marrow.

- **Volatile organic compounds (VOCs).** Commonly found in paint and shown to damage the liver, kidneys, and central nervous system. Without question, choose low- or no-VOC paints.

While a blacklist of toxic chemicals is necessary, please be aware that many of these toxins are being replaced (for market share reasons) with unproven alternatives. For instance, the new BPA-free plastic containers may be no better than the ones with BPA. That means the responsibility is on us to stay vigilant, assume synthetic chemicals are bad for us, and look up specific exposures with resources available, such as from the Environmental Working Group and the U.S. Green Building Council. It is surprisingly easy to minimize the effect of these toxins in our day-to-day life by doing a little research and switching some of your old, toxic products for new, healthier versions. We must demand change with our dollars and our voice.

Various chemical exposures might also be associated with greater breast cancer risk, although epidemiologic studies are mixed.[10]

- The plastic lining in cans (xenoestrogen bisphenol A)

- Flame retardants found in furniture and building materials (polybrominated diphenyl ethers)

- PCBs or polychlorinated biphenyls, used in electrical equipment until banned in 1979 (but still found in imported fabrics)

- DDT (an insecticide used to battle mosquitoes and malaria until banned in 1972; responsible for killing off the bald eagle population)

- By-products of paper pulp bleaching and found in tampons (dioxins or dioxin-like compounds)

While experts quibble about your risk from ingesting synthetic chemicals, I can't emphasize enough the exposure that's mostly within your control: ALAN. I covered this topic briefly in chapter 6 on sleep, but it bears repeating: when you disrupt your delicate inner clock, by reading a tablet until late at night, for example, it may increase the expression of your cancer genes, particularly breast cancer.[11] So start with protecting your circadian rhythm and production of melatonin.

Alternative Beauty and Skin Products

We all use beauty products to try to smooth and clear the skin with age. Here are the alternatives that I use daily.

Skin care. Choose organic if you can for your cleanser, moisturizer, serum, oils, and makeup. I use the Healthy Living app (powered by the Skin Deep database) from the Environmental Working Group and consult it when ordering skin-care products online. My favorite brands, tried and true, are listed in the Resources section.

Hair care. I got my first set of highlights in my thirties—and loved them. Then my middle sister, Anna, explained that I needed something called lowlights (a dark, permanent hair dye), so I got those too. Then a friend told me her life was forever changed by a chemical hair straightener (no more flatiron)—and I realized it was time to look at the risk of these common hair-care products. Little did I know that dark permanent hair dyes contain coal-tar ingredients, which are known human carcinogens, according to the International Agency for Research on Cancer and the National Toxicology Program. Europe has banned these carcinogenic ingredients commonly found in hair dye: aminophenol, diaminobenzene, and phenylenediamine. Yet the FDA here in the United States continues to permit them. Why? Because in the United States, synthetic chemicals, including carcinogens, are still innocent until proven guilty, and absolute guilt is hard to prove.

Meanwhile, the science mounts against hair dye: 23 percent greater risk of breast cancer.[12] Limited reports link permanent hair dye to non-Hodgkin's lymphoma, multiple myeloma, acute leukemia, and bladder cancer.[13] Despite that finding, another study found no such association.[14] Perhaps the reason for the mixed results is the dose, in which case you'd expect hairdressers, who are around hair dye more than the rest of us, to have higher rates of cancer. Indeed, they do: 27 percent higher risk of lung cancer, 30 percent greater risk of bladder cancer, and 62 percent increased risk of multiple myeloma.[15] Bottom line: avoid hair dye (see safer alternatives later in the chapter and Resources).

Mystic Grandma

Need a facial and maybe a rehab for your negative stereotype of aging? Go see Deborah. She is a sixty-five-year-old facialist, hairstylist, makeup artist, yoga teacher, and nutritionist. Women in Marin County see her to bring out their beauty, regardless of age. She grew up unconventionally in the Mission District of San Francisco, where her Nicaraguan mother would take her to a *curandera* (a native healer or shaman) for healing. Deborah has always been immersed in a culture that was comfortable with aging and death. When I ask her about aging, she responds: "I have peace about it. I know that if I die tomorrow, I have a peaceful life. I'm not afraid of death. Our body goes through a cycle, the cycle of nature. Yet many of my clients fear aging and view it as deterioration."

Deborah isn't deteriorating. She wears almost no makeup; she's lean and works out hard with her trainer, performing high-intensity interval training four times per week. She meditates for at least thirty minutes every day. She looks like a woman in her forties. During our conversation, Deborah leans forward and confides that often the women who sit in her chair at the salon are suffering because of their signs of aging and feel like they are not as attractive as they used to be. Sometimes they even ask her to turn the chair so they don't have to look in the mirror. But Deborah has a gift for unearthing beauty.

What does she tell these women? "You have these different stages of life and that beauty is cultivated within. Beauty as you age comes through your eyes, spirit, and outlook on life. That's what I learned from my mentors. My grandchildren want to hang out with me. They bring their friends from college: 'You've got to meet my grandma to see how she takes care of herself, how she eats and exercises . . .' " Deborah embodies ageless beauty, and young people want to be around it because it's rare and uplifting.

The Air We Breathe

Our homes include a number of toxic culprits besides the ones I've mentioned that often go unnoticed and unaddressed. When my husband, a pioneer in green building, starts talking about air quality, my eyes cross. I can't help it. Granted, there's an urgent need to discuss and clean up the air we breathe in our homes. Here are a few reasons why.

You breathe about three thousand gallons of air each day. If the air quality is poor, your health will suffer in more ways than you may realize. In addition, air pollution harms the planet. Air quality has been slowly improving since passage of the Clean Air Act in 1970,[16] but you must remain aware of the latest harms so that you can protect yourself and those you love. For instance, the massive methane gas leak in the Los Angeles area has effected the health of livestock and possibly humans, yet very

Ozone

Ozone is one of those things most people have heard of but can only vaguely define, like gluten. Here are the facts: Ozone, a component of smog, is a gas molecule that consists of three oxygen atoms bound together. It's a good thing when it's high in the upper atmosphere, where it shields us from the sun's radiation. But ozone air pollution at ground level aggressively attacks lung tissue when you breathe it. How does it develop? Ozone is created when the gases that come out of your car's tailpipe and your local smokestacks interact with sunlight, creating a reaction of hydrocarbons, VOCs, carbon monoxide, and nitrogen oxides. The main problem is burning of fossil fuels, such as gas, coal, and oil. Exposure to smog shortens your life span, and women and children are especially vulnerable.[19]

few major news outlets reported the story.[17] Environmental activist Erin Brockovich called the methane leak the worst U.S. environmental disaster since British Petroleum's oil spill.[18]

At first, breathing polluted air may cause your eyes and nose to burn. It contributes to asthma, which affects thirty million adults and kids in the United States. We need to save the air, because ozone and fine particles can hurt your body, irritating your airways and making you cough and wheeze, decreasing your lung function by turning your poor lungs red and swollen, increasing blood clotting and risk of cardiovascular disease, causing developmental and reproductive harm, making you more susceptible to infection, triggering skin cancer and cataracts, exacerbating asthma, and sooner or later making you die prematurely.[20]

Pollution may make you hungry and weaken your bones. Kids in Mexico City exposed to pollution show significant changes in their biomarkers: high concentrations of PM2.5 (a common pollutant in the air) raise leptin levels and lead to vitamin D deficiency.[21]

Now, you can't get away from pollution, but you can make lifestyle changes that will better help your body cope with the pollution around you. Additionally, prevent pollution by improving your air at home, buying products that are no- or low-VOCs, and opening your windows to let in fresh air.

Mold

As an example of the exposome in action, let's get moldy. This past year, I performed a lot of genetic testing on myself and was unhappy to find that I'm a member of a club I didn't want to join: the one in four unfortunate humans who have the mold susceptibility genes. *Blech!*

You might not be able to see or smell it, yet mold may be growing in your home and could be the reason you feel sick. Water damage begets mold, and the genes that make you susceptible to mold illness are the same genes that control your sensitivity to other issues, such as a propensity for collagen breakdown in your skin (the foundation that keeps your skin firm and plump), allergies, plus yeast infections and an inability to process alcohol well, which leads to the formation of acetaldehyde, a toxin.

It can be difficult to diagnose someone with mold toxicity because it can mimic many other conditions. Symptoms are nonspecific and include:

- Memory problems, brain fog, trouble with focus and executive function

- Fatigue, weakness, postexercise malaise and fatigue

- Muscle cramping, aches and pains, joint pain without inflammatory arthritis, persistent nerve pain, "ice-pick" pain

- Numbness and tingling

- Headache

- Light sensitivity, red eyes, and/or blurred vision

- Sinus problems, cough, shortness of breath, air hunger, asthma-like symptoms

- Tremors

- Vertigo

- Persistent nerve pain

- Abdominal pain, nausea, diarrhea, appetite changes

- Metallic taste

- Weight-loss resistance

- Night sweats or other problems with temperature regulation

- Excessive thirst

- Increased urination

- Static "shocks"

Mold can grow in your bathroom, in the showerhead, or in the corner near your shower, especially if the room is not well ventilated. We found visible mold under a bathroom sink where a pipe was leaking. Mold can attach to your shoes, pets, clothes, carpets, furniture, books, and papers. Mold can circulate in your air system, especially if you're like me and rarely change your filters. (The recommendation is to change HVAC filters once every one to three months.) Water-damaged buildings create a complicated mixture of contaminants present in the air and dust, resulting in a toxic chemical stew.

Mold illness from water-damaged buildings is a serious health problem. Unluckily, I have the bad HLA genes that regulate my immune response to mold and other biotoxins. If you're in the lucky group, the 75 percent of people who aren't mold-susceptible, here's what happens when you walk into a house with an active water leak and mold: you inhale mold spores and toxins, and your immune system attacks it successfully by making antibodies. But if you're one of the unlucky ones, like me, without the protection of antibodies, the toxins get recirculated. Most of these folks don't know that they have a genetic susceptibility. The illness is built into their DNA and, once triggered, the inflammatory response and resulting symptoms can last for years and will continue to provoke illness unless it's treated.

If you suspect you may have a problem with mold, start with the Resources in the appendix to find a health professional who can determine if you're genetically susceptible. Then work with your specialist to test your home and other places where you spend significant time.

How Your Liver and Kidneys Cope with Toxins

Your liver is like the chemical treatment plant of your body, and it's often at a loss about what to do when barraged with chemicals from the skin, airways, blood, and the gastrointestinal tract. Your body is designed to flush out toxins, and you have some amazing organs that function solely to remove toxins. Yet when you are overexposed to chemical toxins or trauma, your liver and kidneys work overtime, leading to accelerated aging and illness from a backup of unprocessed toxins. If too much of a backup accumulates, you start to feel more symptoms.

It's important to know about the basics of how your liver detoxifies chemicals because then you can rescue it when it gets overwhelmed. In my

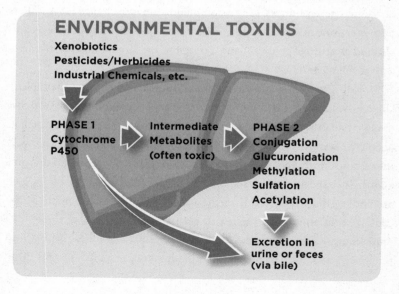

previous book, *The Hormone Reset Diet*, I used an analogy that I learned from a friend to explain the complexity of liver detoxification: your trusty organ the liver is your body's natural filter, designed to purify the blood and remove toxins.

Your liver does this in two phases: garbage generation (phase one) and garbage collection (phase two). In phase one, your liver takes toxins, like BPA, out of your blood and converts them into molecules known as metabolites. In phase two, your liver sends the toxic metabolites to your urine or stool. In other words, you take out the garbage, like I do every Sunday night at my home.

Lamentably, most of us have a problem with both phases. From stress and constant exposure to toxins, you may have an overactive phase one and create too much garbage—some of which is worse than the original toxin itself. Then, to make matters worse, you forget to collect the garbage by neglecting your body's need for detox. It keeps piling up, as if the garbage collectors were on strike. The result is that your liver isn't doing its job of detoxification, which can lead to the symptoms of toxic exposure. By increasing your intake of key minerals, fiber, and other nutrients, you can strengthen the garbage collection and removal capacity of the liver.

You'll learn more in the protocol about the food and supplements that support phase one and two of liver detoxification.

Mitochondria Are Your Toxin Warriors

To make energy in the body, your mitochondria convert fat and other fuels into energy your body can readily use. Regrettably, your mitochondria take a lot of hits along the way and may not be able to accomplish this important task. If there's a gap between the energy your body needs and your mitochondria's ability to fulfill it, you will feel tired, maybe even toxic.

Your mitochondria look mighty when healthy, but they're actually quite delicate when it comes to harm from environmental exposures. There are several causes of mitochondrial dysfunction. One is nutritional deficiencies, which occurs when you're not getting the antioxidants that

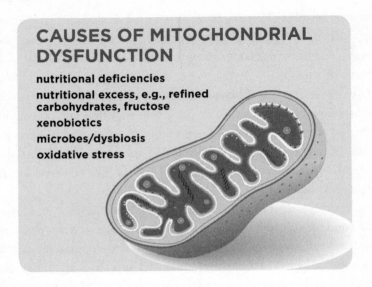

CAUSES OF MITOCHONDRIAL DYSFUNCTION

nutritional deficiencies

nutritional excess, e.g., refined carbohydrates, fructose

xenobiotics

microbes/dysbiosis

oxidative stress

you need from green tea, fruits, and vegetables to offset free radicals. Another is nutritional excess, such as when you eat or drink too much sugar. Xenobiotics, synthetic chemicals, and endocrine disrupters further damage the mitochondria, as does microbes, altered gut flora, and direct oxidative stress (in addition to poor intake of antioxidants). The result is that you feel tired and can't fully cope with the daily assaults of life.

Starting this week, you'll have a plan to perk up your mitochondria and get them working for you again. You'll increase your consumption of fruits and vegetables and take supplements that help reset mitochondria directly. You'll reduce your body burden of biotoxins that make your mitochondria unable to power your body. In sum, you'll begin to feel like yourself again, starting deep within your cells.

Ways to Counter the Stress of Environmental Toxins

Now that you're totally depressed, please know that there are ways to reduce and remove negative exposures and ways to expose yourself positively—that is, to counter the stress of environmental toxins. First, swap out toxic skin and home products for safe ones. For example, start using stainless-steel or glass containers for food storage. Don't use plastic to heat food. And cook with pans that are not coated in plastic. (See Resources.) Second, incorporate positive exposures: saunas, cruciferous vegetables, fruits, nuts, green tea, and supplements.

Sauna Exposure: Warm Your Heart and Longevity Genes

It's hot to get hot. Improve your exposome with positive exposures such as saunas (dry saunas and infrared) or heat, (hot tubs or steam rooms). Of all these forms of heat, dry saunas have the most evidence that they help you age well, but infrared saunas are not far behind. If you want to live long and healthy, you need molecular chaperones to tend to your DNA, and that's what sitting in a sauna provides.

Sauna is a heat stressor, a form of hormesis that resets the body, including the DNA. It's like a bench press for your longevity genes, but without the grunting. When you sit in a sauna, you activate the longevity gene called FOXO3, which turns on the genes for stress resilience, antioxidant production, protein maintenance, DNA repair (prevents mutations), and tumor killing. Most of these genes decrease their genomic expression with age.

In addition to turning on other important genes, FOXO3 makes heat-shock proteins. Heat-shock proteins ensure that the proteins in your body are properly folded, like a fitted sheet, not bunched up and wrinkled. Poorly folded proteins clump together and cause damage (such as in atherosclerosis, congestive heart failure, and neurodegenerative diseases like Alzheimer's), leading to shorter lives. Heat-shock proteins also neutralize oxidative stress, which acts like rust in your body. You won't be surprised to learn that when you make more FOXO3, you triple your chance of living to one hundred.[22]

In a study published in the *Journal of the American Medical Association,* researchers found that men who enjoyed a sauna four to seven times per week had a 40 percent drop in mortality of all causes![23] So the sauna may be the best place for your heart as you age.

We all have senescent cells in the body, which are like zombies—not quite dead, but not quite alive. They secrete pro-inflammatory cytokines that damage nearby cells. FOXO3 is like zombie patrol: it turns on the genes of autophagy, a fancy word for programmed cell death. FOXO3 also modulates immune function so that your inner police can control the bad guys: bad bacteria, viruses, and cancer cells.

Contraindications to sauna bathing include unstable chest pain (angina), recent heart attack, and severe aortic stenosis. Sauna bathing leads to an increase in heart rate and reduction in total vascular resistance, which consequently lowers blood pressure. Talk to your health-care professional if you have any of these conditions, but keep in mind the relative safety of saunas. In one study from Finland where sauna bathing is common, the annual death rate occurring while in a sauna was less than 2 per 100,000.[24]

I grew up taking Finnish saunas with relatives in Minnesota, followed by a jump in a cold lake. Steam is created by throwing water on hot rocks heated by wood fires. Finnish saunas heat you from the outside in, while infrared saunas heat you from the inside out using infrared waves. Infrared waves penetrates about two inches deeper than Finnish saunas, allowing water molecules (about 70 percent of your body) to vibrate and raise your core body temperature so that you sweat.

If, for some weird reason, you're not yet convinced that you need to schedule in the sauna this week to sweat away your toxins, here are other proven benefits:

- Improves athletic performance (endurance, less muscle atrophy, more plasma volume)[25]

- Enhances heart rate variability[26] and balance in the nervous system

- Increases insulin sensitivity (more glucose receptors on muscle cells), resulting in a striking reduction of 1 percent unit in the glycated hemoglobin, fasting glucose, and body weight in people with diabetes[27]

- Generates "runner's high" and boosts growth hormone and testosterone[28]

- May correct your lipids[29]

- Increases capacity for stress tolerance by reducing the sympathetic nervous system, as measured by reduced adrenaline and cortisol in just seven days[30]

The bonus is that sauna bathing is relaxing; it eases stress while adding to your healthspan. Start with twenty minutes four times this week.

Other Exposome Weapons: Cruciferous Vegetables, Fruits, Nuts, Green Tea

Another positive exposure that you can easily fit into your daily life is eating more of certain foods that detoxify your body.

Cruciferous vegetables. Most important, increase your intake of cruciferous vegetables: broccoli, Brussels sprouts, cabbage, cauliflower, kale, bok choy, watercress, and other similar vegetables. The main way that exposures enter your body is through your skin and the lining of your gastrointestinal system. Seventy percent of your immune system is beneath the layer of your gut, which is thinner than a piece of tissue paper. Vegetables trigger cleanup in your immune system.[31] In fact, the very same receptor on cells that environmental toxins use for their bad xenobiotic effects is also used by cruciferous vegetables. So by eating more cruciferous vegetables, you crowd out the bad environmental toxins.[32] You may prevent cancer. You dose yourself with sulforaphane, which inhibits phase one and stimulates phase two in the liver. You receive fiber to purify your liver, and vitamin C to counteract free radicals. Kale alone increases your antibody production five times your normal antibody count.[33]

Fruits and nuts. When I first moved to San Francisco for my residency training in obstetrics and gynecology, my grandfather remarked that it was the "land of fruits and nuts." Offended at first, I had to chuckle when my grandpa came to my wedding and afterward had to navigate around the Gay Pride parade to get back to the airport on Sunday. But Grandpa was onto something—I soon realized the importance of fruits and nuts to mitigate toxic exposures. In fact, fruits and nuts provide some

Your Vision, Aging, and Exposures

Just as physical strength drops with age, so does eye strength, or vision, because the eye's lens stiffens with age. As mentioned, most eye doctors consider presbyopia, meaning, literally, "old eyes," to be an unavoidable aspect of aging. I disagree. If you consider the exposures involving your eyes and what you can do to modify them, you just might prevent or reverse old eyes. The exposure here is the near reading people do on their laptops and smartphones. The problem is that very few people know about this exposure and what's proven to help. Before you surrender to reading glasses dangling grandma-style around your neck (or LASIK, multifocal eyeglasses, or multifocal contact lenses), consider these preventive measures:

1. **Check.** Have your vision checked regularly by your eye professional. It's important to track your progress over time, both at home and in the eye professional's office. Watch for danger signs such as headaches, blurred vision, and eye strain.

2. **Eat for young eyes.** Keep eating the Younger food plan, as described in chapter 5. Remember to continue consuming plenty of vegetables, fruits, and fish because they contain the nutrients (vitamins, minerals, healthy fats, antioxidants) that can help slow the aging of the eye.

3. **Manage medical conditions** that can affect the aging of your eye. Keep fasting blood sugar 70 to 85 mg/dL. Prevent autoimmune conditions by treating your immune system well; problems such as rheumatoid arthritis put you at greater risk of old eyes. That means avoiding inflammatory foods and unmitigated stress. Finally, keep your thyroid function in top shape (read my first book, *The Hormone Cure*, for more information).

4. **Wear sunglasses** to block ultraviolet rays when you're outdoors.

5. **Palm.** Gently relax and massage your eye muscles daily this week. It's common to feel eye strain from near work on the laptop and smartphone. Here is one of my favorites, called palming. Your eye muscles can become fatigued just like muscles elsewhere, but we rarely think to relax them actively. Rub your hands together vigorously, then place you warm hands on your eyes to soothe and relax the muscles. Leave them there, applying gentle pressure for about one minute. Palming relaxes strained eyes. When you spend a lot of time doing up-close work, your eye muscles can get locked into tension patterns and lose the ability to focus at different distances.

6. **Box it in.** To help relax your eye muscles, imagine a box. Now look up to the right corner, inhale and exhale, then look up to the left corner, breathe, and repeat for all four corners. Do this daily.

7. **Do near/far exercise.** When you're performing close work, like I am now on my laptop, take a periodic break. I keep a pencil by my side for this exercise and sit facing a window with a view of the distant landscape. Hold the pencil about eighteen to twenty inches away from your face. Look at the pencil point and then slowly pull it to the bridge of your nose. Repeat three times. Then, holding the pencil in front of you, look to the horizon of your view outside. Trace the horizon for five seconds or more, then focus again on the pencil point. Repeat three times. This helps your eye muscles focus your lens over near and far distances, which will help keep your lens young.

These eye exercises can help to compensate for the overdevelopment of some eye muscles and may help prevent presbyopia.

of the strongest evidence supporting the interaction between genes and nutrients.

Most people know that nuts are nutritionally good, but few know that they reduce oxidative stress. Nuts are like an anti-rust treatment in your body. Walnuts are proven to reduce oxidative stress.[34] Eating one Brazil nut per day increases selenium and glutathione production, and glutathione is the most powerful antioxidant your body makes.[35] That allows you to run your thyroid more efficiently and detoxify chemicals better.

You already know that vegetables pack an antioxidant punch. What about fruit? My favorites are berries (blueberries, blackberries, strawberries, raspberries), plums, and citrus (oranges, lemons, limes). According to the Human Nutrition Research Center on Aging at Tufts University, these fruits have the highest measure of antioxidant power as measured by ORAC (oxygen radical absorbance capacity). Blueberries, as an example, have more antioxidants than forty other fruits. One cup of wild blueberries gives you more than thirteen thousand antioxidants, about tenfold higher than the USDA's measly daily recommendation. I tend to avoid fruit juices and dried fruits like prunes and raisins because of the concentrated fructose, which may cause fructose overload, leading to insulin resistance, fatty liver, and high blood pressure, as I wrote about in *The Hormone Reset Diet.*

Green tea. A morning cup of green tea not only wakes you up without the usual rev of caffeine but also provides a rich dose of polyphenols, antioxidants that prevent oxidative stress from making you old and sick, and it inhibits phase one while stimulating phase two of liver detoxification. It also stimulates an important transcription factor called Nrf2, which regulates oxidative damage. One of the shining stars of green tea's cast of characters is epigallocatechin gallate (EGCG). Green tea leaves, stems, and buds contain six types of antioxidants known as catechins (of which EGCG is one), and all six clear your body of free radicals and weaken the cold virus.[36] That means when you're exposed to the cold virus, you're less likely to get sick. Taken further, studies show that green tea helps prevent and reverse liver disease, infections, digestive issues, heart disease, neurodegenerative diseases

(Alzheimer's, Parkinson's), hormonally related cancers (breast, endometrial, ovarian, prostate), all-cause mortality, and even genital warts.[37] On a practical level, you can brew green tea or take it as a capsule; my recommendation is to become a master at brewing it.

Fruits, nuts, and green tea offer clear examples of nutrigenomics, or the interactions between your individual genetic makeup and dietary components that result in modulation of genetic expression.

Protocol for Week 5: Expose

You may be wondering if it's even worth trying to limit your exposure to synthetic chemicals, pollution, and mold. They seem to be everywhere. Thankfully, it's easy to build up your defenses. After all, you've already done the hard work of feeding your genes, getting more sleep, exercising right, and releasing habitual patterns of tension. Plus you've been eating green vegetables at least twice per day, cooking more at home, and finishing your last meal of the day at least three hours before bedtime.

Here is your daily template for week 5. For the next seven days, follow these guidelines as closely as possible and tune into subtle changes that occur.

Basic Rituals

Food

- Wake up, brew, and drink at least one cup of green tea each day.

- Increase your daily intake to nine to eleven servings of fruits and vegetables. An easy way to get this done is to make a shake and add one to two servings (one or more cups) of greens. This adds two to four servings at a time, which is very efficient!

- Add broccoli sprouts to your menu; eat one cup per day. This counts as one of your servings and provides more sulforaphane for phase one and phase two support in the liver.

- Eat a minimum of one Brazil nut or three shelled walnuts per day.

- Lengthen your overnight fast to sixteen to eighteen hours once or twice this week (once for becoming younger, twice for weight loss). For instance, finish dinner by 6:00 and fast until noon the next day. When you fast for a longer interval overnight, inflammation clears up and you may reduce your risk of breast cancer.[38]

Beauty

- Swap your skin care and cosmetics for an organic brand such as Annmarie Skin Care and Tarte. Download the Environmental Working Group's app Healthy Living (powered by their Skin Deep database) to your smartphone (it's free!).

- I'm a fan of applying organic antioxidant serum to the face after cleansing and then applying an organic oil to seal the deal. Alpha-lipoic acid (5 percent) has been shown to reduce aging in the face over twelve weeks.[39]

- For nail polish, I recommend one of the following brands, which have a score of 2 or less on the EWG's Skin Deep rating system.

  - Zoya is my favorite and scores a 1, or the lowest toxicity rating

  - Acquarella also scores a 1

  - Keeki Pure and Simple scores a 2

  - Keeki Pure and Simple Base Coat and Top Coat is another better choice if you choose a toxic brand of polish over the base coat

- For covering gray hair, I recommend Hairprint, the remarkable discovery of Dr. John Warner, a chemist in Massachusetts. He developed a nontoxic and safe way to mimic the normal function of hair follicles; namely, to infuse hair with natural pigment, which works only for black or brown hair. If you have gray hair, it restores the innate color in eighty minutes. Yet the Hairprint is so safe, you can eat it. All that's happening is that you replenish your hair's natural pigments, which restores the health of the hair, resulting in stronger hair with greater body and luster. It's a healing system for hair.

Supplements

My patients and my husband are always asking me for the easy way—they want supplements they can take so they avoid the work of all the other changes in the Younger protocol. My answer is that it's not so easy if you want to get it right.

You might think it's a matter of simple math: if you have more oxidative stress, just counteract it by taking antioxidants in pill form. Not quite. It turns out that eating healthy food comes first. Studies that isolate an antioxidant, such as beta-carotene, are mixed in terms of how it affects health. It's also not physiologic to simply isolate one nutrient. And even more concerning, antioxidants can flip into oxidants (by giving up an electron, thereby becoming a free radical).

Whole foods are a better choice. Many nutrients that you get in food form, like the proanthocyanidins from berries, you can't get as a supplement. So I'm repeating here for the record: *whole, organic food first* whenever possible. Then support and enhance your diet with supplements, but only when you are eating the right foods first. Supplements cannot fix a nutrient-poor diet.

That said, this week the focus will be on mitochondria-loving and energy-boosting supplements, like alpha-lipoic acid (ALA). Even with a whole-foods diet, it's hard to get enough to keep your oxidant/antioxidant status in balance. ALA repairs damaged cells, and it's one of the most critical antiaging, anti-inflammatory, and antioxidant agents you can ingest (or apply to your skin—see Beauty section).[40] ALA is four-hundred times stronger than vitamins C and E. It occurs naturally in mitochondria, but you need to take 300 to 1,800 mg per day for it to act like a free radical warrior in your body.[41] ALA may protect your bones as you age[42] and keeps

Air Filters

Invest in a good air filter to remove particulates such as dust, mold, pollen, and microbes from circulating air so that you're less likely to suffer from allergies, asthma, and mold-related illness. The best air filter is IQAir. You have two options: portable units (HealthPro Plus) spaced throughout home, or a whole-house filter (Perfect 16) that connects to your HVAC system. The Perfect 16 is more powerful because it moves a lot more air and ends up being more affordable if you're going to buy more than three portable units.

your cells sensitive to insulin so that your blood sugar doesn't climb.[43] One study in obese women showed that ALA aids weight loss when they are on an energy-restricted diet.[44] Other studies found a benefit at 800 mg/day for weight loss,[45] but yet another study confirmed that 1,800 mg is superior to 1,200 mg for weight loss in women and men.[46]

Home

- Consider testing your tap water for toxic ingredients. A thorough study by the Natural Resources Defense Council in nineteen urban areas found several toxic contaminants, including rocket fuel (perchlorate, a thyroid disrupter and potential carcinogen), lead, and arsenic.[47] See Advanced Projects for more details.

- Swap your laundry detergent for a natural one. Take a look at your other cleaning products; switch to organic or make your own.

- Remove plastic food containers and replace with glass or stainless steel. Use only glass or ceramic dishes to heat food in the microwave.

- Discard plastic-coated (e.g., Teflon) cookware and use cast iron or enamel-coated cast iron.

- Visually inspect for mold in the house—under sinks and around the bathtub. Use a mixture of water and vinegar to remove. Seek help from a professional if significant (more than a rim around the tub).

Advanced Projects

- Purchase an air filter for your home (see sidebar).

- Test your tap water for toxic contaminants. Purchase a home filtration system as indicated.

- Grow your own broccoli sprouts (see instructions in appendix).

- Have your home inspected for mold.

- Avoid heavy metals; swap your tuna for salmon to reduce mercury exposure. Remove dental amalgams. Wear organic lipstick to prevent lead exposure.

- Test your liver. Measure your ALT either through your health professional (best) or on your own (see Resources for recommended labs for self-ordering).

- When you're eating out at a restaurant with unknown food quality or are drinking nonorganic wine, pop two activated charcoal pills to block some of the chemicals and toxins you may ingest. I keep a bottle in my purse and take 500 to 600 mg before ingesting the food or drink.

Recap: Benefits of Week 5

It can seem overwhelming to learn about all the exposures that are bombarding you every day. Don't stress; your body is already designed to contain toxins. You just need to help it along. Once you do, you may experience:

- Lower oxidative stress and more antioxidants, so you can tilt your aging equation for the better

- Improved collagen and connective tissue, so you wrinkle less, starting today

- Better, healthier, maybe longer telomeres

- Longer healthspan

- A greener, healthier home environment

- A diet that combats cancer with a whole-foods approach and supports phase one and two of liver detoxification

Bottom Line

This week you're preserving your youth by reducing and mitigating exposures. Toxins can disorganize your body inwardly, starting with your mitochondria. I believe that aging accelerates in your mitochondria, so getting your power factories back on the right path is crucial. When you heal your mitochondria by preventing exposures, clearing out the damage with antioxidants, and eating the right foods, you are activating your age-suppressor genes. You are inwardly reorganizing, like you might do once per year in your closet so that getting dressed is easier. It bears repeating that environment is responsible for 90 percent of your risk of disease, so altering your environment is key. Beauty, diet, and home exposures have the biggest influence on your health—and these three are the easiest for you to alter and control.

CHAPTER 10

Soothe

WEEK 6

There is a place in the soul that neither time, nor
space, nor no created thing can touch.

—Meister Eckehart

I startle easily. I shut my eyes during violent films. My amygdala, the primitive part of the brain that's constantly on the lookout for danger, is hot. I don't clear stress well. It loiters in my body like a juvenile delinquent, poking holes in my gut wall, ruining digestion, truncating my timekeeping telomeres, and smacking around my poor defenseless brain. Even everyday stress—driving to out-of-town volleyball tournaments, visiting relatives, finding clothes to wear to an important event, treating lice—can make me feel ridiculously vulnerable, traumatized, and tense. It's as if the factory forgot to install my shock absorbers, so I've had to come up with my own.

I know from my functional medicine practice that my experience isn't unique. Yes, everybody knows plenty about stress management, yet as I've mentioned, common knowledge isn't the same thing as common practice. Most people keep trying the same old ineffective strategies. The first part of any health-related change is to understand why the old way is problematic; the second is to alter the old way, especially fixed behavioral patterns.

Successful coping requires skillful monitoring of the stress-response system and an ability to shut down the response when stress is over. Put another way, we want to flip the switch from the trance of fight-or-flight to the more evolved state of tend-and-befriend.

This chapter is about soothing the emotional tension, whereas chapter 8 was about releasing the physical tension. The goal isn't just relaxing more; instead, we want to go after the genes responsible for making you feel so stressed and frozen, identifying the genetic variations and working with them in a different and more productive way. Then make it stick. It's one of the biggest challenges to our evolution as a species, because chronic stress makes you prematurely old.

Eventually, I learned how to turn stress into an ally—and not just by bathing in Epsom salts by the ton. Here are the key genes that most commonly affect stress levels; we can turn them off or on to help release stress so we can function better daily and build resilience.

- FKBP5, or FK506 binding protein 5, a gene that plays a role in the stress-response system in the body, the hypothalamic-pituitary-adrenal (HPA) axis

- CYP1A2, cytochrome P450 family 1, subfamily A, polypeptide 2, the gene that codes for an enzyme that makes you more likely to overstimulate the adrenal glands (and metabolize caffeine slowly)

- FAAH, fatty acid amide hydrolase, also known as the Bliss gene, controls the enzyme that acts on anandamide, our natural cannabinoid molecule of bliss. Derived from the Sanskrit word *ananda*, or "bliss," when anandamide binds to the cannabinoid receptor, you feel calm and happy. I need more anandamide, and you may too!

- WWC1, WW domain-containing protein 1, the gene that codes for the KIBRA protein and involves memory and synaptic plasticity

- MR, the gene controlling one of the regulators of the HPA, the mineralocorticoid receptor, which programs us to have increased

ACTH (the hormone that signals your adrenals to produce more cortisol) and cortisol in response to psychological stress. When a neighbor was robbed in his home at gunpoint, I lost my mind and planned a move to a safer community. It took me weeks to calm down. My husband began researching German shepherds online and bought a Taser, although I suspect he secretly wanted an excuse to buy one.

- TH, the gene that codes for tyrosine hydroxylase, makes the body freak out in a cold environment. I can't go surfing in the cold ocean or sit in a bathtub of ice without feeling incredibly stressed. I have to live in a warm place or else my cats, short for catecholamines (stress neurotransmitters and hormones), and blood pressure rise too high. (It's the same gene associated with white-coat hypertension, a common stress response that can age you prematurely.)

When it comes to stress, modern life is at odds with our ancient genome. We are hardwired to survive a rare threat and then chill out for a few months before the next crisis. Instead, psychological, emotional, or work-related crises pummel most of us daily, hammering our cells with excess stress hormones and shortening healthspan. I've tried many things over the years to clear stress: exercise, yoga, transcendental meditation, more time with girlfriends, mindfulness, chanting, chi running, orgasmic meditation, pole dancing. But let's face the facts: we're not designed to be constantly or even frequently stressed. We must program regular downtime to relax, unplug, slow down, and digest life. I'll guide you toward one of several options to improve your shock absorbers and, thereby, slow down the clock.

Why It Matters

If you're like me and have difficulty soothing yourself, please know that you're not doomed to be anxious, burned out, overly sensitive, in survival mode, or depressed for the rest of your life. Epigenetics is a powerful tool to allow you to create the downstream reality of your choice. You can adapt for the better. It requires that you start to name stress and its many guises—feel it, track it, hunt it, and slow down your reaction. The process asks something big of you: you have to broaden your repertoire. Epigenetic change, like you've been making throughout the Younger protocol, can be empowering or disempowering for your body as a whole. Until you shift to being on the offensive or are already extremely skillful in responding to stress, your epigenetic changes will most likely add to your stress load, making you old before your time.

First, the big why: when your relationship with stress is screwed up, you age ten years faster.[1] Things get stuck in your craw more often than you'd like—such as last night when my upstairs neighbors at a hotel were loud and boisterous past midnight, and security didn't hear a problem! You run around with your fight-or-flight system (that is, your sympathetic nervous system) stuck in the on position, which gives you physical and psychic wrinkles (your inner skin or proteins get poorly creased and folded), blood-sugar problems, brain shrinkage in the hippocampus (the area responsible for memory consolidation and emotional regulation), and thins your bones. You can't tune in to others or what you most cherish. In other words, you can't turn on your mirror neurons—the neurons that discharge when an individual performs an action and when he observes another individual perform an action, a key process of belonging and connectedness.[2] Mirror neurons are the seat of self-awareness. They're central to associative learning and more sophisticated behavior. Some consider mirror neurons the most important discovery of the past decade in neuroscience.[3] Instead, you feel stuck, separate, self-centered, fearful. The aperture of awareness narrows, and healthspan declines. You may feel it as time compression, judgment, aversion, or irritability. See the figure for further details about the risks of unbounded stress.

Second, if those issues don't grab your attention and motivate you to change, consider this: *traumatic stress has lasting effects on your body and on the bodies of your kids and grandkids via epigenetic change.* Your level of stress and how you perceive it may not change your hardware (genes), but it can change your software (epigenetics). Your negative perceptions

YOUR BODY
STRESSED

BRAIN
Disrupted circadian rhythms, increased risk of depression and dementia

EARS
Ringing (tinnitus), hearing loss

SLEEP
Decreased deep sleep (phase 3); insomnia

BELLY
Increased fat deposits

GUT
Acid reflux, decreased blood flow, leaky gut and food sensitivities, altered transit time, irritable bowel syndrome

OVARIES
Lower sex hormones, diminished sex drive

BONES
Reduced bone density, increasing fracture risk

SKIN
Increased oil production, acne, eczema, psoriasis, cold sores, wrinkles

MUSCLES
Spasm, chronic muscle aches and pains

EYES
Blurred vision, fatigue, pain in and around the eyes, headaches, twitching

THYROID
Lower production of active thyroid hormone, T3

HEART
Increased heart rate and blood pressure, higher risk of heart attack and stroke

ADRENAL GLANDS
Cortisol and adrenaline surges, lower sex hormones, glucocorticoid resistance

LIVER
Blood sugar surges, pre-diabetes, diabetes, metabolic syndrome

IMMUNE SYSTEM
Inflammation, more susceptibility to infections and autoimmune conditions

CELLS
Shortens telomeres, the timekeeping caps on chromosomes

BLOOD VESSELS
Stiffens arteries and raises blood pressure

of life events lodge themselves in your body molecularly, affecting your DNA, like a soul wound.

Rachel Yehuda, Ph.D., knows about soul wounds because she's studied epigenetic changes in the survivors of the Holocaust and the 9/11 terrorist attacks. She is a professor of psychiatry and neurosciences at Mount Sinai Hospital. First, she looked at the genes of thirty-two Jewish survivors of the Holocaust. Then she looked at the genes of their children. She compared their results with Jewish families living outside of Europe during the war. Her focus was FKBP5, the gene that regulates your stress-command center.

She found epigenetic inheritance of survivor trauma; the survivors' DNA did not change, but the epigenetic marks did, and those changes (the sticky notes attached to FKBP5 and an increased risk of PTSD) were passed on to the survivors' offspring.[4] Yehuda found further evidence of inherited problems with FKBP5 in women who were pregnant and at or near the World Trade Center in New York City during the 2001 terrorist attacks. In a group of thirty-five pregnant women, alterations to FKBP5 increased a woman's risk of developing PTSD and passing it on to her baby.[5]

Before anyone found evidence of transgenerational fear in humans, researchers found it in mice. In the original studies of the epigenetics of trauma, mice received electric shocks while they were smelling cherry blossoms. The juxtaposition caused the mice to fear cherry blossoms. Researchers found that their babies and grandbabies were afraid of cherry blossoms too. Trauma was inherited, not in the genome itself, but in the

The Other Half of the Icarus Myth

Remember the Icarians from chapter 4? Their name comes from the Greek myth of Icarus, which most people remember only half of. Icarus is the guy whose father made him wings out of feathers and wax so he could fly. Maybe you recall from the story that it's bad to fly too close to the sun, which is what Icarus did, thereby melting the wax, losing his feathers, and falling to his death. Perhaps the way you live now, indulging stress and needing a glass of wine to take off the edge, is equivalent to flying too close to the sun. But in the Icarus myth, there's another problem besides flying too high; Icarus was also told not to fly too low, because the mist of the sea could weigh down the feathers of Icarus's wings. We need to avoid doing both. Don't run the sympathetic nervous system too high or too long, and, conversely, don't run it too low with low cortisol and under-reacting to threat. Longevity comes from the broadest and most adaptive dynamic range.

epigenome, or collective marks (like sticky notes), that can tell genes what to do.[6]

The takeaway? When I get lazy about clearing stress from my body or being consistent about my meditation practice, it helps to think of my kids and how I'm wired to pass on my trauma response to them. So if you can't wrangle stress for your own benefit, do it for your future generations.

Science of Week 6: Soothe

People who perceive the world to be stressful not only look haggard, but also develop allergies and other problems with their immune system. They wheeze and itch more, and they die early of heart disease.[7] Our collective need to self-soothe is growing. You'd think by now we'd understand how stress gets under the skin, but the truth is that we've only recently figured it out.

Studies of twins can help us understand the old nature-versus-nurture question (genes versus environment). In one study of three hundred sets of twins, investigators asked what affected levels of job stress more, the day-to-day environment or the individual's personality.[8] First, differences in personality type are determined almost half by genes. Second, when it comes to work stress, genes affect 32 percent of the variance of job stress from one person to another and almost half of the variance in health problems. That's more than the $^{90}/_{10}$ rule that governs your risk of serious disease, so working around the genetic determinants of stress is very important to your job satisfaction and health.

The Normal Stress System

In the normal stress response, a threat (for example, a deadline, your babysitter canceling, a bear ambling toward you) triggers the *hypothalamus* part of your brain to jump into action. The hypothalamus is the intersection of your nervous system and hormone system. It's part of a group of structures, called the limbic system, that interpret threat. In the end, the hypothalamus regulates the following: body temperature, hunger, sleep, emotion, and other homeostatic (balanced) systems. Stress disrupts homeostasis, evoking a reaction in the nervous and endocrine systems to respond to the stress and then return to homeostasis or a new level of balance.[9]

Here is the normal cycle of the stress-response system:

- The hypothalamus secretes corticotropin-releasing hormone (CRH) and arginine vasopressin (AVP) in response to stressful threat.

- CRH signals the pituitary to make adrenocorticotropic hormone (ACTH), which stimulates the adrenal glands to make stress hormones, including cortisol. CRH can also be released outside of the central nervous system—for example, in the skin—where it may cause inflammation.[10]

- Cortisol travels back to the brain. In the hypothalamus, cortisol turns off the production of CRH and AVP. In the rest of the limbic system, the cortisol turns off the glucocorticoid receptor (and stops the signal to make more cortisol) and turns on the mineralocorticoid receptor (keep making AVP).

Going back to my husband's response to the armed robbery in our neighborhood, his stress hormones went through these pathways, he decided to buy a weapon and campaign for a guard dog, and that was the end of the story for him. Balanced state achieved. For me, the experience was very different and prolonged.

How Stress Can Hurt

The other thing that happens when you perceive stress is that glucocorticoids are released into the blood so that you can run or fight. This is powered by raising your blood pressure, heart rate, and blood sugar. If this happens every once in a while, say every three to six months, it's normal, and the body adjusts accordingly right after. But if you have genes that program you to anticipate stress, perceive a high level of stress, or recover slowly and/or poorly, excess stress hormones may become toxic to your system. High levels of glucocorticoids shorten your telomeres, which may arrest some cells into the zombie-like state of senescence (where the cell is neither alive nor dead) and release chemical messengers that promote inflammation.

Most studies suggest that the imbalance that makes the HPA axis hyperactive occurs in people with anxiety, depression, or PTSD. Even more interesting, the heightened stress reactivity seems to predate the mental diagnosis and may be associated with certain genetic variations of the key stress genes (mentioned at the beginning of the chapter: FKBP5, CYP1A2, FAAH, WWC1, MR, TH),[11] presenting health-care providers and citizen scientists with a golden opportunity to intervene before all the bad stuff takes over. The biggest problem in dysregulation of the HPA axis is a failure to suppress CRH, AVP, and ACTH, meaning that the stress response doesn't get turned off properly. The result is that you keep feeling stressed even when the threat is gone, like I did for weeks after the crime a few blocks from our home.

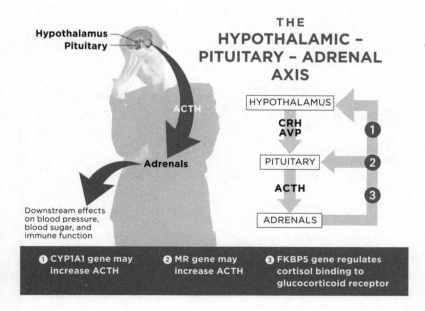

THE HYPOTHALAMIC – PITUITARY – ADRENAL AXIS

Hypothalamus
Pituitary

ACTH

Adrenals

Downstream effects on blood pressure, blood sugar, and immune function

HYPOTHALAMUS

CRH
AVP

1

PITUITARY

2

ACTH

3

ADRENALS

1 CYP1A1 gene may increase ACTH

2 MR gene may increase ACTH

3 FKBP5 gene regulates cortisol binding to glucocorticoid receptor

An amped-up perception of stress causes lower vagal tone (or responsiveness), which means the vagus nerve is not fully performing its functions. Your vagus nerve is the most important nerve in your parasympathetic nervous system. If the vagus nerve isn't happy, you won't be healthy and are more likely to age faster. Vagus means "wanderer"; the nerve wanders all over your body to important organs such as the brain, neck, ears, tongue, heart, lungs, stomach, intestines, liver, pancreas, gallbladder, kidney, spleen, and reproductive organs in women. Lower vagal tone is linked to a variety of problems:

- Anxiety

- Poor satiety or sense of relaxation while eating

- Difficulty accessing mind-body connection and flow state

- Low stomach acid secretion

- Poor absorption of B_{12}

- Low or slow bile acid production, so it's harder to clear fats and toxins

- Constipation

- Poor blood flow to kidneys

- Higher blood pressure

- Poor glucose control

- Poor heart rate variability and greater risk of heart disease

- High resting heart rate

- Frequent urination

- Limited or absent capacity for orgasms

Conversely, high vagal tone is a marker of greater altruistic behavior and closeness to others.

Get the Right Dose of Exercise

We're all aware that inactivity and sitting too much are bad, but just like Icarus's mandate not to fly too high or too low, there's a middle ground with exercise that provides the greatest longevity benefits. When you don't exercise enough, it can harm your immune system, reduce your stress resilience, and dysregulate your circadian rhythm. When you exercise too much—too long, too intensely, too frequently, and without sufficient recovery—you may cause problems to your stress-response system, leading to immune problems, injury, and a leaky gut.[12]

Overexercising releases two key hormones: CRH and cortisol. CRH increases the permeability or leakiness of the intestinal wall as well as the permeability of the lungs, skin, and blood-brain barrier. Cortisol levels rise with rigorous exercise, such as running, which may cause too much wear and tear and accelerated aging.[13] High cortisol also alters tight junctions between cells such that small harmful substances may pass through the barrier. Additionally, high cortisol reduces gut motility, blocks digestion, blunts blood flow to the gut, and lessens mucus production, an important immune function. Elite athletes get help from several work-arounds, such as by supplementing with probiotics,[14] omega-3s,[15] and vitamin C;[16] however, modulation may be your best bet.

Five Ways People Tend to Deal with Stress

According to revered psychotherapist and Buddhist teacher Sylvia Boorstein, we are born with five default ways of coping with pain or tension. Here are the five responses to stress:

1. Fretting

2. Getting angry

3. Losing heart and feeling defeated

4. Personalizing (*It's me*, or *It's my fault; I did something wrong.*)

5. Searching for sensual soothing, like pizza or a doughnut or sex

These five coping methods arise naturally, like factory settings in your body and mind. If you know your default coping mechanism, you don't have to let it define and victimize you. Let yourself off the hook. Instead of making yourself more miserable by trying not to be this way—anxious, angry, eating a pint of ice cream—accept it as your innate programming. There's something mysterious about acceptance; it soothes an overactive nervous system. Acceptance is the gateway to working with tension more wisely.

It's not so much the realities themselves that are a problem when it comes to traumatic stress and accelerated aging; it's the way we struggle against them and keep reliving them. Life inevitably brings disappointments and complications, but we can modify our response and perspective. As Sylvia Boorstein soothingly suggests, "Sweetheart, you are in pain. Relax, take a breath, and let's pay attention to what is happening, then we'll figure out what to do."

I know you know this: suffering often results from resisting, rather than accepting, issues beyond your control. Remind yourself that we can change only what's within our control. It's the common sense that almost nobody follows. Acceptance has many faces. It's the healthier alternative to trying to escape or gorge or avoid reality; these may be temporarily soothing, but long-term health truly requires acceptance and active surrender. It may mean forgiving the person who fills you with rage, saying no when you mean it, or deciding to eat what's on your plate and not go back for seconds.

Best Ways to Ride Stress

We all have an innate capacity to counteract toxic stress. Herbert Benson calls it the relaxation response. When you add a new approach to your default response, that's the start to retraining your mind and responding to stress differently. You can get there through prayer, mindfulness, yoga, qigong, and many other modalities.

By now you know that I love yoga and use it as a way to create my own shock absorbers. The relaxation mechanism of yoga was described twenty-five hundred years ago in the Yoga Sutras of Patañjali: *Yoga citta vritti nirodhah* (Yoga is the ending of disturbances of the mind).[17] When you synchronize your breath with your movement, your mind slows down and enters a more settled, aware state.

Meditation does it too. In a study published in 2009, forty-four subjects (half meditators, half nonmeditators) underwent brain scans (high-resolution MRIs). Meditators had more gray matter in the parts of the brain responsible for attention, mental flexibility, mindfulness, and emotional control.[18] Other studies have confirmed the brain's improvements with the ability to focus mindfully.[19] Meditation retrains the brain and mind structurally, physiologically, and psychologically so that you navigate stress better.

Meditation will stimulate your vagus nerve, but you can improve vagal tone in other ways too. Maybe it's going to church or synagogue. Perhaps your version of church is the gym or knitting. To paraphrase author Elisa Albert, the main point is that stress is going to drag you down if you don't learn how to ride it.[20] It's like labor and delivery. Work with me here (as suggested by Albert): *climb on and ride it.*

Seven Ways to Help Your Vagus Reset

1. Connect positively with others[21]

2. Take a cold shower (try it, but if it stresses you out, you may have the same TH gene as me!)[22]

3. Schedule a reflexology (foot massage) session[23]

4. Sleep on your right side[24]

5. Sing![25]

6. Get acupuncture, especially in the ear[26]

7. Book a craniosacral session (described in chapter 8)[27]

These actions may trigger your stress genes to turn off, bringing a greater sense of calmness.

Finding the Right Way

As a gynecologist who has counseled women for many years about how to handle their genetic tendencies and out-of-whack hormones, I know that stress is a causal or exacerbating factor in nearly every hormone imbalance. I take it as a given that most people would benefit from overhauling their relationship with stress. The path begins with you objectively observing your experience in reaction to stress. Then you can find the best avenue for meditation, which is my top choice for transforming how you turn off your stress response.

Meditation is well documented to help regulate stress, anxiety, chronic pain, and many illnesses.[28] Meditation improves positive mood and may help you resist the flu. Benefits begin immediately after training, even in beginners.[29] The important part is to make it a daily practice starting this week.

Just as there's no single path to nirvana, there's no single best way to meditate. The four main ways are:

- Focused attention (focus placed on a single object, such as the breath or a visualization—examples are loving-kindness and centering prayer)

- Open monitoring (rather than focusing attention on a single object, you openly monitor all aspects of your experience without judgment—examples are mindfulness and vipassana)

- Transcendental meditation (a type of meditation where you repeat a word or seed mantra)

- Movement meditation (yoga, labyrinth walking, walking meditation, orgasmic meditation)

Each modality for soothing oneself fits certain personalities better than others, but all of them help switch your genes to a different mode. For me, yoga works the best because it helps me witness my thoughts rather than be ravaged by them. Furthermore, I can't sit quietly until after I move first. Yoga may not be your preferred method to reset your relationship to stressors, so here's the quick guide to identify the modality that fits you best.

Dr. Sara's Modality: What's Best for You?

1. If you're completely new to meditation, start with *open monitoring*. Listen to a Jon Kabat-Zinn–guided mindfulness meditation, or download a podcast from Tara Brach.

2. If you're a type A, hard-driving, restless, agro type of a person who cannot stop thinking obsessively, I suggest that you try *transcendental meditation* or *movement meditation*. You probably won't do well in a slow-moving Iyengar or Hatha yoga class. I recommend vinyasa, Ashtanga, power, or Forrest yoga. The goal is to observe your thoughts rather than abolish all thought. If you're a sensualist, try *orgasmic meditation*, a sequenced practice in which one partner gently strokes the other partner's clitoris for thirteen minutes. The result is considered to be therapeutic rather than sexual, presumably because the stroking releases a flood of oxytocin.

3. If you're hard on yourself or able to hyperfocus without difficulty, try *focused attention*. If you believe in God, *pray*. Loving-kindness in particular is an excellent place to start because it involves focusing on warm, compassionate thoughts about yourself and others. Sylvia Boorstein is one of my favorite teachers of loving-kindness (also known as metta). Here's one of her loving-kindness meditations to focus your thoughts:

   > May I feel protected and safe.
   > May I feel content and pleased.
   > May my physical body provide me with strength.
   > May my life unfold smoothly with ease.

4. If you're an addict, which tends to be transmitted across generations, try a *Twelve Step program*, such as Alcoholics

Anonymous, Food Addicts, Overeaters Anonymous, or Co-Dependents Anonymous. The foundation of recovery is to develop a relationship with a higher power and to work the twelve steps, a road map for personal integrity. Twelve Step programs are built upon regular meeting attendance and "working the steps" with a sponsor or group.

There are more forms of meditation, but most tend to be derived from one of these four. Various forms may enhance vagal tone but the three proven modalities are loving-kindness, chanting "Om," and tai chi.[30] Find out what works for you, and if you're still unsure, try each of these methods for one day and choose the one you like best for the rest of the week and the remainder of our seven-week protocol. (I'm terrible at focused attention and open monitoring. Trying to eat half a cookie and feel joyous about it sets me up for failure.) The goal of all of these methods is to give the mind a warm bath and reset cortisol levels.

Close Encounters with Transcendental Meditation

It was 1984, and most girls my age were singing lyrics to Madonna's music. Instead, I was seated in a large lecture hall learning transcendental meditation. Someone famously said that people won't remember your words but they'll remember how you made them feel. I don't remember the words of our instructor, who looked like a crusty sadhu from India, but I certainly recall how the practice made me feel: peaceful, content, at ease.

I had done yoga with my mother and great-grandmother while growing up, but I didn't have a consistent practice at that time or understand that the purpose of yoga was to still the mind. I learned that the human brain takes life's inevitable complications and hardships and makes them worse with unnecessary suffering. I discovered how my mind creates conflict with my experience: *I want this, not that; I'm cold and need to get out of here; why did my nail polish just chip off?* Transcendental meditation provided an important gift: I found the gap between my old distress pattern and a wiser way of being. In that gap, my behavior and actions started to change.

As you search for the right fit, set a goal of short duration—about ten minutes—so that it's an easy win. If you can't do ten minutes, start with five. If ten minutes is easy, try twenty. There's a benefit to simply sitting with erect posture and breathing more deeply, even if your mind is monkeying around. You are still altering your stress response and getting on the path to retraining your mind.

Protocol for Week 6: Soothe

It's time to make stress your ally so I'm going to make the protocol really easy and stress-free this week. Pick the meditation modality you want to try and consistently practice every day. I suggest that you meditate immediately upon awakening. This will give you the greatest chance of success, because you will start retraining your mind before the stress of the day hits you.

When you meditate, practice breathing deeply in and out through the nose. Take full-lung-capacity breaths as a way to dial down the sympathetic nervous system and dial up the parasympathetic nervous system. You need to turn on the serratus posterior superior muscle to breathe deeply (to do this, let the belly fully expand as you inhale) and turn on the serratus posterior inferior muscle to empty your lungs fully (to do this, let the belly fully contract toward your spine). Both muscle actions are needed to kick on the parasympathetic nervous system. One or both of these muscles may be tight from underuse. When you get these muscles working for you, taking a deep inhale and exhale becomes easier over time and creates healing in the body and mind.

Even though exercise should already be part of your daily routine during this protocol, if you have not been consistent, use this week to get back on track. Choose an exercise that you enjoy and break a sweat four times this week. As we know, exercise combats stress, helps us sleep better, and raises endorphins.

Besides seeking to eliminate stress as much as possible in this week's protocol, address your reaction to stress. Figure out your default coping mechanism and combat stress with a new relaxation response. I suggest you be as mindful as possible throughout the day, and when you encounter a stressful situation, take three deep breaths before you react. Over the span of the week you will notice how much more present and in control you are as you face daily challenges.

This week the focus is on regular meditation (any type that works for you) in order to change your response to stress. Long-term you'll make your brain healthier, more flexible, and more agile.[31]

Basic Rituals

- After waking up, meditate, pray, or listen to a guided visualization for ten minutes each day. For smartphone resources, see Resources at the end of the book.

- Choose one approach to raise your vagal tone as listed in the "Seven Ways to Help Your Vagus Reset" (see page 221). It may take more than a week for you to notice a higher vagal tone, but moving forward, continue to incorporate these resets into your life whenever possible.

- Go to bed by ten o'clock, lying on your right side to stimulate the vagus nerve. Do your best to secure seven to eight and a half hours of sleep nightly (see chapter 6 for a refresher).

Supplements

- Take omega 3, such as extra virgin cod liver oil or other fish oil, 1 to 2 grams per day. It lowers your cortisol levels, increases lean body mass, and improves vagal tone as measured via heart rate variability.[32]

Advanced Projects

- If you need help to stay accountable with your new practice, find someone to work with you. Sometimes a life coach or therapist can hasten your progress, assist you in identifying your difficulties with stress, and offer suggestions to handle stress differently. Or consider doing this protocol with a close friend so you can help each other.

- Centering Prayer. This is my simplified, agnostic version of the practice popularized by others.[33]

  - Select a sacred word that symbolizes your connection to inner divinity. Examples: *grace, calm, trust, faith, peace, ease.*

  - Sit comfortably, and close your eyes. Silently use your sacred word to invite the connection to the divine.

  - When you find yourself becoming attached to thoughts, return to the sacred word, a symbol of your consent.

  - Remain still with eyes closed for a few more moments.

Recap: Benefits of Week 6

When you make stress an ally instead of being victimized by it, you immediately improve your physiology by reducing the damage of your stress response. This is where we begin to actually retrain the brain and stop accelerated aging. Starting to meditate immediately alters your mind-set. In as little as a few minutes of deep, diaphragmatic breathing, you will note a sense of calmness, which arises from the immediate blood flow changes within the body. The less stress you feel, the calmer your amygdala, the part of your brain that looks for threats.[34] You'll turn on your mirror neurons and, consequently, your self-awareness and introspection. The more consistently you practice meditation, the easier it will be for you to maintain mental control over your responses to daily stress. Remember to choose a modality of meditation that best fits your personality so that it can become a daily part of your routine.

In addition, when you add consistent meditation, your digestion improves, your immune system will function better so you're less likely to catch the flu, and you may improve your telomeres, which may reflect slower aging. Have you ever noticed how people who practice meditation seem to look more youthful? There is something to this ancient practice, and science now substantiates this belief.

Perhaps most critical to me as I learn to ride stress better is that when I'm not jacked up on stress, I am available to love more and better. Arne Garborg put it beautifully: "To love a person is to learn the song that is in their heart and to sing it to them when they have forgotten." I can't sing my loved one's song without my mirror neurons and a steady sense of calm.

Bottom Line

When you know what's going on within your body in the midst of stress and become aware of the downstream risks, you can use meditation to change old patterns and adapt better. Instead of resisting what you don't like or don't want, try cooperating with the way life is emerging in this very moment—like the song says, "If you can't be with the one you love, love the one you're with." Take what comes, and you'll live longer and happier.

CHAPTER 11

Think
WEEK 7

Almost all our suffering is the product of our thoughts. We spend nearly every moment of our lives lost in thought, and hostage to the character of those thoughts. You can break this spell, but it takes training just like it takes training to defend yourself against a physical assault.

—Sam Harris

Your head is where your brain and mind reside, although most of us give our minds more attention. The health of each affects the other, so we don't want to neglect either. In fact, the perception of your external environment affects the biology of your internal environment, and vice versa. Your thoughts can direct gene expression. So we want to integrate the mind and body, starting with the brain.

This goal is born of necessity because the brain is getting older. I like to think it's getting better, because I know that the math of a healthy brain involves a better balance between inputs (like drinking green smoothies and meditating) and outputs (such as prioritizing sleep, preventing leaky gut, and even keeping mitochondria in top shape). Part of the challenge for me is that I have a high level of negative self-talk, cognitive dissonance, and distortion, plus a familial tendency toward addictive behavior. If something is worth doing, isn't doing more of it better? *Uh, no, not usually.* I'm a recovering workaholic, exercise-aholic, self-fixer-aholic, to-do-list-aholic, and food addict. Or as a friend put it, I'm a type A minus.

Maybe you have one or more of these problems too. You almost certainly experience negative and inconsistent thought patterns, attitudes, and beliefs that influence your choices and behaviors, sometimes irrationally and without your conscious awareness. You may be like me in the way that I overvalue the wrong factors when making decisions, default to an all-or-nothing mentality, or dwell on the negative rather than the positive. While it's totally human to be slow to acknowledge errors, I know that these tendencies have led to conflict in my marriage, self-righteousness in my parenting and friendships, and sometimes to eating and drinking outside of my personal integrity.

There is an alternative. Each of us can think and act with more awareness, generally doing a better job at this opportunity called life. It requires mental rehabitulation and a recalibration toward what is good for the mind, brain, and nervous system. Without all of those operating at their best function, your choices for a longer healthspan suffer along with your inner clock—that is, you get old before your time. It's not as hard as you might think, as long as you break it down into baby steps. What you've accomplished in the previous six weeks of the protocol is already benefiting your head; now we want to take the improvements to the final level. At this point, you won't be surprised to see me rattle off the top genes that give a person stinkin' thinkin' or hurt your brain function. I am genetically loaded with some, but thankfully not all, of these variations.

- Corporate warrior gene, or COMT, which affects whether you're a warrior or a worrier (shockingly, I'm a warrior).

- Alzheimer's and Bad Heart gene, APOE, which raises or lowers your risk of Alzheimer's disease (I don't carry the APOE4 risk allele, but many of my patients do); the changes start in the brain decades before symptoms appear.

- Unhelpfully named adrenergic beta-2 surface receptor gene, ADRB2, which gives me a terrible relationship with weight and food over the long haul. When you have this variation, weight loss takes you twice as long as it takes a normal person. It's like a party crasher for women who want to burn fat. Exercise helps very little; it's all about wrangling my mind to eat the right types and amounts of food with steady discipline. *Gah!*

- Klotho, a gene that codes for an antiaging hormone that raises IQ.

- Dopamine receptor D2 (DRD2), which makes me more likely to overeat and behave addictively.

- BDNF, the gene that makes your brain smarter and more neuroplastic (adaptable) over time as you get older.

- FAAH, also known as the Bliss gene, the same one we covered in the last chapter, which codes for the enzyme that acts on anandamide, our natural cannabinoid molecule of bliss.

An extreme example of faulty thinking is Alzheimer's, a brain disease that affects your mind. Symptomatic of the disease, victims have a total lack of awareness that they're losing their minds. Two-thirds of Alzheimer's patients don't understand that they have it, and my grandmother Helen was a good example. Granny stood out as a beacon of unconditional love when I was young—glamorous, funny, always humming and dancing to big-band songs while she puttered around her home, cooking, cleaning, and showering me with care. I'd clomp around her house wearing her spectator pumps and costume jewelry. While I was growing up in Maryland, she picked me up after school (and camp during the summers) because my mother worked full-time. Granny taught me how to garden, fish, and catch blue crabs. That is, until she developed the early stages of Alzheimer's in her sixties.

First, Granny would get lost while driving. We'd go grocery shopping and end up on the highway to Baltimore instead of heading toward her home on the Chesapeake Bay. Then she became less articulate and was no longer able to trust her instincts; she developed a vacant stare, like no one was home. Later, her warm and loving personality changed; she would glare at me with fury and probably frustration as her mind slipped away. The smallest decision would render her childlike, paralyzed by pathos, a ghost of her former self. Yet she didn't die. Instead, she languished in a nursing home for twenty more years, unable to recognize anyone in our family, bedbound and requiring round-the-clock care until her death at age eighty-four. She had a long life span but a painfully short healthspan.

I've learned that developing Alzheimer's may be avoidable. Even if you're genetically wired to develop it based on your family history or genes, you can prevent or reverse this heartbreaking scourge. I don't have the gene for Alzheimer's, but given my experience with my grandmother, I'm extremely motivated to do everything within my power epigenetically to prevent the disease from taking hold.

Why It Matters

Normally, the mind and brain age gradually, if at all, in the absence of injury and disease. When you take care of your brain, vocabulary and language skills can actually increase with age. Older folks may be better at perspective and problem solving; they take their time to carefully craft smart and well-reasoned solutions to problems. They are better at recognizing patterns and being attuned to the effects of their decisions.

We all know elders who are wiser and more accepting of the mixture of joy and sadness in their lives. They are less likely to feel angry, stressed, or worried than younger people. Psychological wellness—roughly, your overall appraisal of your life and mood—reaches a low point at age forty-six.[1] After that, with the brain and mind in a better place, the body benefits from improved stress resilience and wiser choices. Gratitude, forgiveness, appreciation, and a sense of calm peak at age seventy and stay high. Other functions—organization, planning, and analysis—are the same in elderly brains as they are in youthful ones. However, it isn't all rainbows and unicorns. Processing speed tends to slow, and memory becomes weaker in the aging brain. As collagen recedes, joints give out, and more gray hairs keep appearing, these tough situations may breed negative thoughts and depression if you let them get to your head.

My hope is that this won't happen to you because you're already on a better track with our protocol! You can prevent deteriorating memory and dementia (and hearing loss!) through regular physical exercise, learning and education, repairing your gut, and even playing video games. You want your hippocampus—the section of your brain that serves as the center of emotion, memory, and the autonomic nervous system—to be lush and connected, like a dense fern growing in a tropical rain forest.

A Simple Way to Create New Neural Pathways

Here is a simple technique based on an exercise from the late Irish poet and philosopher John O'Donohue[2] that I've adapted with a few more ideas from meditation teacher Tara Brach.[3]

1. Document. Take a day this week to recognize and document your self-talk. Sometimes I call my thoughts the "members of the committee" who sit in judgment inside my head. Just record the most common thoughts without judgment.

   - *I'm fat.*

   - *I'm so tired. It's not normal to feel this tired.*

   - *I'm getting old. My neck hurts.*

   - *I'm a bad mom.*

   - *I deserve chocolate.*

   - *I suck at getting together with my friends.*

   - *I need to call my mom; why am I so bad at calling her regularly?*

   - *Look at that woman in the front row of barre class! She's got twenty years on me and she's stronger! I'll never be like her!*

   No editing. Simply write down the top five to seven thoughts.

2. Ask. For each thought, inquire: Is it true? Is it helpful? Is it serving me? What's the tone? How old is the narrator? The idea is to be a dispassionate observer of your thoughts, so that you're not fused with a limited set of emotions. For instance, maybe I do suck at getting together with my friends, and I need to address it. Note: swear words are a sure sign that the narrator is a rebellious teenager!

3. Identify the bad, but with kindness. Does it make me feel bad? Explore the downstream emotional consequences of the self-talk. For me, thinking that I'm fat becomes a self-fulfilling prophecy. It's not helpful or loving, and it keeps me constricted and stuck.

4. Recognize the good. What's good about the thought? When it comes to Sylvia in the front row of barre class (see chapter 4 for a complete description), comparing myself to her causes me despair, but drawing inspiration and learning from her is great. She is a wonderful model of optimal aging.

5. Reframe. Is there a new way to frame the self-talk so it's more loving and supportive? As John O'Donohue described in an interview with Krista Tippett,[4] you've been married to these thoughts for a long time, and you've never even flirted with other thoughts. Now's the time to flirt with a cognitive reframe that suits you—your mind, brain, emotions, spirit—better. Such as *Look at Sylvia. She is strong and beautiful. I want to age like her. I think I'll take a nap today like she does, skip the restaurant for dinner, and make sure I get to bed by ten.*

Science of Week 7: Think

If you want to skip the science, go straight to the protocol to start changing your brain for the better.

When it comes to the ways that your brain ages and slows down, there are four main pathways in your brain cells: inflammation, mitochondrial dysfunction, intracellular calcium overload, and oxidative stress.[5]

These molecular events, which often overlap, become toxic to nerve cells in the brain. What's a girl to do? Your job is to prevent all of these molecular problems or, if just one or two are happening, address those to keep the brain happy, adaptable, and slow to age. I'll show you how in this week's protocol.

1. Inflammation. Of course, inflammation is a natural part of your body's defense system, but problems arise when it doesn't turn off—for example, when the immune system is chronically on high alert. In the introduction, I referred to this process as *inflammaging* because unnecessary inflammation accelerates your aging, like a woodstove that keeps burning until the house is torched. Persistent immune activation leads to the flood of noxious chemicals in the body that can trigger neurodegeneration, as seen in Alzheimer's and Parkinson's disease.[6] When biomarkers of inflammation are elevated, cognitive decline is more likely.[7]

2. Calcium overload. Normally, calcium levels rapidly rise and fall in a cell as a means of triggering biochemical signals, such as the release of neurotransmitters. You need normal calcium signaling to stay neuroplastic, or adaptive. Calcium and its movements become disturbed in several neurodegenerative disorders, such as Alzheimer's, Parkinson's disease, Huntington's disease, and in the motoneuron disorder amyotrophic lateral sclerosis (ALS, also known as Lou Gehrig's disease). When calcium levels get disturbed, they can change gene expression, harm the mitochondria (there's a recurrent theme here and lots of overlapping), reduce neural plasticity, and cause problems with neuron survival.[8] Sadly, even small changes in calcium levels can wield big changes in cognitive function.[9]

3. Oxidative stress. As we covered in chapter 9, oxidative stress is the imbalance between harmful chemicals (free radicals like reactive oxygen species and H_2O_2) and neutralizing antioxidants (such as glutathione). Oxidative stress is a major player in brain

fog and accelerated aging. When you have too many free radicals interacting with your genes, immune system, and endocrine system, this stress generates that foggy feeling. Your hypothalamus, the part of your brain that makes many essential hormones, is most influenced by oxidative stress. This stress accumulates like rust and causes mitochondrial dysfunction, which then leads to more oxidative stress and increased inflammation . . . you can see the vicious cycle.[10]

4. Mitochondrial dysfunction. When you're healthy, your mitochondria power your thoughts and actions. Millions of these tiny organelles in your cells work together as a power grid for your body, a bioenergetics hub of energy. Your mitochondria lose their might from various causes that usually overlap: nutritional deficiencies like neutralizing antioxidants, nutritional excesses such as carbohydrates and/or fructose, bad microbes and dysbiosis, xenobiotics (particularly pesticides, herbicides, and substances such as hydrogen sulfide), abnormal mitochondrial DNA, and excess

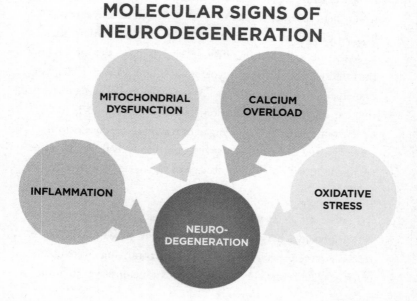

MOLECULAR SIGNS OF NEURODEGENERATION

MITOCHONDRIAL DYSFUNCTION

CALCIUM OVERLOAD

INFLAMMATION

OXIDATIVE STRESS

NEURO-DEGENERATION

oxidative stress. When your mitochondria don't work, you age fast throughout your entire body.[11] You feel tired and worn out, often because your mitochondria are tired and worn out.[12]

Our Biggest Fear: Alzheimer's Disease

Most people consider old age as a progressive decline toward drooling and living in a nursing home, which is one of our greatest fears. By 2050, the number of people age sixty-five and older with the big "A" is expected to have tripled, according to conservative estimates.[13] Sadly, after age sixty-five, an individual's risk of developing Alzheimer's doubles every five years. After you hit age eighty-five, the risk reaches nearly 50 percent.[14]

Alzheimer's is characterized by a decline in memory, language, problem solving, and cognitive ability. It stems from a loss of certain nerve cells and an accumulation of beta-amyloid plaques and neurofibrillary tangles. The plaques result from abnormal protein folding and aggregation, like the misfolded bedsheets I described in chapter 9. Two-thirds of people with Alzheimer's are women. The best known gene for it is APOE4, which has the strongest impact on your risk. You inherit a copy of the APOE gene—e2, e3, or e4—from each parent.

When you consider most diseases of aging, the issue is usually a problem of imbalance. Take osteoporosis, for example. With this disease, your bones are on the wrong side of the balance; you have too much osteoclast activity (which thins the bone) and not enough osteoblast activity

Greek Myth, Klotho, and Oxidative Stress

Your genes can shift how fast your brain is aging. For instance, the Klotho gene codes for the Klotho protein, which protects cells and tissues from oxidative stress and therefore acts as an age-suppressor gene.[15] Klotho takes its name rather poetically from Greek mythology; Klotho is Zeus's daughter who spins the thread of life.[16] Making more Klotho—with exercise and enough vitamin D—regardless of the gene you inherited will help you keep spinning your thread of life and prevent it from being cut too early.

(which builds the bone). A similar problem is occurring in the brains of people with Alzheimer's, according to neurologist and UCLA professor Dale Bredesen of the Buck Institute for Research on Aging. In the normal brain, certain signals create more nerve connections and make memories, while other signals foster a way to forget irrelevant information, like when you need to do some spring cleaning in your closet to make room for new clothes. In people with Alzheimer's disease, the balance between opposing signals is off, leading to a net effect of truncated nerve connections (synapses) and lost memories of important information. If you are freaking out and wondering if there's a solution, yes, almost there!

You Can Reverse Cognitive Decline

The latest version of the Alzheimer's Facts and Figures gets one fact completely wrong: "It's the only cause of death in the top 10 in America that cannot be prevented, cured, or slowed."[17] Since first described a century ago, Alzheimer's disease has been without effective treatment. Until now. Dr. Bredesen has pioneered a program that reverses memory loss in nearly all of his patients within three to six months. Yes, *reverses*. Larger clinical trials need to be done, but this is a rare bright spot in the treatment of Alzheimer's that you need to know about now, before it's too late.

Yet the cure is probably not a single drug with one target. Instead, the best solution appears to be a functional medicine approach that addresses multiple root causes. Imagine having a roof with thirty-six holes in it, and a drug that patches only one hole. Dr. Bredesen says that if you seal one hole, you still have a leaky roof with thirty-five other holes. So taking a drug for treatment isn't helpful. But if you address multiple holes, you may get an additive or even synergistic effect, even if each hole is only modestly affected. You might reduce the leakiness by 90 percent. You haven't fixed everything, but you're much better off.

The Best APP for Your Brain . . . Amyloid Precursor Protein

There's a protein you should know about when it comes to saving your mind: amyloid precursor protein (APP). According to Dr. Bredesen,

APP is like the CFO of your company (that is, your body). "APP looks at all the inputs from all the different accountants: are you in the red or the black? Every day, you're actively *remembering* what you did this morning or you're actively *forgetting* the seventh song that played on the radio on the way to work yesterday. You have this beautiful balance," he says. One hundred percent of the people with Alzheimer's disease are on the wrong side of that balance for years. He nicknamed the problem *synaptoporosis*.

So Dr. Bredesen set out to understand what's driving the process—what leads to Alzheimer's patients being on the wrong side of the balance, and how we can use that information to prevent Alzheimer's from taking hold, before it's too late. Dr. Bredesen published the results of his initial small

Alzheimer's Disease and APOE Gene by the Numbers

- APOE4 is responsible for approximately 20 percent of Alzheimer's cases.[18]

- Inheriting one copy (from one parent) of the APOE4 gene triples your risk of developing Alzheimer's, while inheriting two copies (one from each parent) of the APOE4 gene makes your risk eight to fifteen times higher.[19] Approximately 2 percent of the U.S. population inherit two copies of e4.[20]

- Women who carry the APOE4 gene are more likely to get Alzheimer's than men.

- Not having the APOE4 gene does not guarantee that you will not get the disease.

- APOE4 causes a profound reduction in SIRT1 and an overproduction of mTOR, two of the longevity genes.[21]

- If you inherit two copies of the APOE2 gene, you have a lower risk of Alzheimer's.

- You can test for this important gene in one of the labs listed in the appendix.

study of a comprehensive functional medicine program to reverse memory loss.[22]

One of his first patients was a sixty-seven-year-old woman with two years of progressive memory loss. He referred to her as Patient Zero. She was considering quitting her demanding job, which involved analyzing data and writing reports. By the time she reached the bottom of a page she was reading, she would need to start again at the top, owing to her poor short-term memory. She began to get disoriented driving, and she mixed up the names of her pets.

Patient Zero's mother had developed similar progressive cognitive decline beginning in her early sixties, had become severely demented, and had died at eighty. When the patient consulted her physician, he told her that she had the same problem as her mother, and he could do nothing about it. He wrote *memory problems* in her chart, and she was therefore turned down when she applied for long-term care. Knowing that there was still no effective treatment and that she couldn't get long-term care, she decided to commit suicide. She called a friend to commiserate, who suggested Dr. Bredesen for a second opinion.

She started Bredesen's program and could adhere to some but not all of the components; she couldn't convince her local doctor to prescribe bioidentical hormones; she stayed off gluten but kept eating other grains like brown rice, and she had a hard time getting more than seven hours of sleep. Still, after only three months, all of her symptoms abated—she was able to navigate familiar roads again without problems, recall phone numbers, and read and retain information. When she got sick with a cold, she stopped the comprehensive program and relapsed, but when she reinstated the program, she felt normal.

Types of Alzheimer's Disease

Dr. Bredesen has found three distinct types of Alzheimer's disease.[23] The first is the "hot" or inflammatory type, most commonly found in people with one or two copies of the APOE4 allele. The second is "cold" or non-inflammatory. Women more commonly fall into the type 2 category. Bredesen has found that those with type 1 or type 2 are able to hang on to their jobs for a while because the main problem is memory and they find work-arounds. For instance, they are dentists who can still drill teeth and physicians who can still listen to hearts as long as they have good assistants to guide them. Type 3 or "toxic" results from exposure to specific toxins—most commonly mold—as a manifestation of chronic inflammatory response (CIRS). It affects younger people and the cortex more broadly. They lose the *cerebral cortex*, which is the folded gray matter in the brain that's important for consciousness. That means they lose old memories, fall apart in terms of daily functioning and stress coping, and must leave their jobs early.

Behaviors That May Harm Brain and Mind (and Increase Your Risk of Alzheimer's)

Although 60 to 80 percent of your risk of Alzheimer's is related to genetic factors, only about half of this risk is through APOE.[24] (Other genes such as APP, can cause Alzheimer's, but they are very rare.[25]) Furthermore, you can modify your risk with fewer toxic exposures and other lifestyle adjustments, many of which we've already introduced in the protocol to date.[26] As we covered in the introduction, the five factors contributing to inflammaging also increase your risk of declining cognition and Alzheimer's.[27] Correcting hormone imbalance takes care of the brain, so it can take care of your mind and, by extension, you.

Other culprits include:

Poor sleep

- How it affects your *brain*: you didn't get the full shampoo by your glymphatic system (see chapter 6), so you feel leaden and toxic. Blood sugar may be higher than normal, leading to brain fog. You may even feel depressed or anxious. In a national poll from 2005, people diagnosed with depression or anxiety were more likely to sleep fewer than six hours per night.[28]

- How it affects your *mind*: you're groggy, cranky, and have a shorter fuse. Cortisol is higher than normal, allowing stress to take advantage of you.

Lack of stimulation

- How it affects your *brain*: What you don't exercise atrophies. It's true for muscles, and it's true for your brain.[29] The stimulation of adult education reduces the risk of dementia by 75 percent.[30] When you stop engaging in cognitively stimulating activities like crossword puzzles, games, baking, gardening, or staying up on current events, your brain can go south. Picking them up again can help reverse mild to moderate symptoms of Alzheimer's.[31] My father-in-law is a great example of ongoing learning: at age eighty-four, he passed multiple Coast Guard exams to become his local flotilla's commander. Go, Ira!

- How it affects your *mind*: While the same study by the Cochrane Collaboration did not find a benefit to mood, it's clear that lifelong mental engagement keeps your mind sharp. Try humor; it improves memory![32]

Lack of community

- How it affects your *brain*: Interacting with people stimulates the brain and keeps you sharp. Stronger social ties are proven to lower blood pressure and boost longevity. Having no social ties is an independent risk factor for cognitive decline. Talking to another person just ten minutes per day improves memory and test scores.[33] The higher the level of social interaction, the greater the cognitive functioning. Researchers found that people with at least five social ties—such as church or social groups, regular visits or phone calls with family and friends—were less likely to experience cognitive decline than those with no social ties.[34]

- How it affects your *mind*: Dealing with other people can be challenging, but it's an important way to create meaning and reduce isolation. In fact, my friend and colleague Mark Hyman, M.D., says, "The power of community to create health is far greater than any physician, clinic, or hospital." I agree. If having a community of accountability partners doubles to triples your weight loss, imagine what a positive community can do for your healthspan! Socializing is just as effective as other types of mental exercise when it comes to improving memory and intellectual performance. Plus it's just *more fun!*

Poor microbiome

- How it affects your *brain*: There's a reason that the gut is considered the second brain. Nine meters long, with five hundred million neurons, featuring thirty major neurotransmitters, it is a chief ally in the function of your nervous system. Problems with the microbiome are associated with autism, anxiety, and depression. On the other hand, a good microbiome, rich in lactobacillus and bifidobacteria, may increase BDNF.[35]

- How it affects your *mind*: I can't put it better than *The Atlantic*: Your gut bacteria want you to eat a cupcake.[36] When your gut microbes and their DNA are out of balance, you may crave sugary foods in order to feed the troublemaking microbes, thereby inducing a vicious cycle of more cravings and more bad microbes.

Thinking Ahead

The best news out of all of this scary stuff is that cognitive decline can be fixed with targeted lifestyle changes, and women have an advantage, if they're not too old, because of the benefit of natural hormone balancing.

Want to know what Patient Zero did? You've already been performing a streamlined version of it in previous weeks, and we'll add a few additional tenets in this week of the protocol.

- Cut out all refined carbohydrates, gluten, and processed and packaged foods.

- Added vegetables, fruit, and wild fish.

- Fasted for three hours between dinner and bedtime, and for a minimum of twelve hours between dinner and breakfast.

- Purchased an electric toothbrush and flosser and used them regularly.

- Started practicing yoga and, ultimately, became a yoga teacher. She practices yoga sixty to ninety minutes a day at least five times per week.

- Added transcendental meditation twice per day for twenty minutes.

- Began taking melatonin at night; sleep went from four or five to seven or eight hours per night. Other supplements: methylcobalamin, 1 mg/day; fish oil, 2,000 mg/day; vitamin D_3, 2,000 IU/day; CoQ10, 200 mg/day.

- Exercised aerobically for thirty to forty-five minutes four to six days per week.

Now age seventy, Patient Zero remains without symptoms of cognitive decline and continues to work full-time, sometimes clocking ten-hour days and traveling internationally. She feels better than she did thirty years ago, and even her libido is high. She continues to eat gluten-free with an occasional glass of red wine.

Reversing or preventing cognitive decline demands a change in diet, exercise, stress, sleep, brain stimulation, and supplements. Patients with Alzheimer's often have poor hygiene, inflammation, insulin resistance, vitamin D abnormalities and hormonal imbalances, and toxic exposures.

Similar to what you've found in the previous weeks of the Younger protocol, during week 7 you'll make the proven changes to modify your genetic or epigenetic risk of Alzheimer's. The key is to intervene before the window closes—ideally within ten years of the first symptoms (like my granny getting lost on familiar roads), when there's still time to reverse the imbalanced signals. Or, even better, way before any symptoms start.

Increase the Good Inputs

You want to begin optimizing the inputs for your brain as early as possible, such as *now*. Don't wait until you're having significant difficulties, because that's been one of the biggest problems in dementia. We could dramatically reduce the global burden of dementia if people would just get going earlier with improving the brain. As Dr. Bredesen indicated, when you address at least thirty-six neurotrophic inputs in aggregate, you can reverse and prevent cognitive decline. They work together to improve signaling in the brain, from calcium to mitochondria to hormones (like estrogen, progesterone, and testosterone) so that you flip APP toward the remembering side of the equation. Think of it as the brain equivalent of preventing osteoporosis; you are preventing *synaptoporosis*. You just need a plan that's proven to make a difference.

Epigenetic effects account for one-third to one-half of cases of Alzheimer's disease.[37] The epigenetic effects result from environmental factors that hasten the onset of dementia: traumatic brain injury, aging, diabetes, hypertension, obesity, sedentary lifestyle, smoking, low educational levels, and stroke.[38]

Now that you're scared stiff and want to prevent Alzheimer's from entering your body, here's the short list of what prevents it.

Summary: Good Inputs to Prevent Alzheimer's

1. Eat whole, unprocessed food. Consume healthy fats like medium-chain triglycerides and omega-3s.

2. Keep blood sugar in the normal range (fasting 70 to 85 mg/dL).

3. Sleep seven to eight and a half hours per night.

4. Exercise regularly and intensely.

5. Practice yoga (or another method of observing your thoughts and calming down your nervous system).

6. Create hormone balance, getting a prescription for bioidentical hormone therapy as needed. (Read my first book, *The Hormone*

Exercise Type, Intensity, Duration

In a study conducted at multiple academic centers (among them the University of Maryland and Cleveland Clinic), a group of ninety-seven cognitively intact older adults, ages sixty-five to eighty-nine, were divided into four groups: high genetic risk (APOE4) and low physical activity; low genetic risk (no APOE4) and low physical activity; high genetic risk and high physical activity; and low genetic risk and high physical activity. Low physical activity applied for people who exercised at low intensity (slow walking, light chores) twice per week or less. High physical activity included one or more of the following three or more days per week:

- Brisk walking, jogging, or swimming for fifteen minutes or more

- Moderately difficult chores for forty-five minutes

- Regular jogging, running, bicycling or swimming for thirty minutes or more

- Playing sports such as tennis for an hour or more

After eighteen months, only the group with high genetic risk and low physical activity showed shrinkage of the hippocampus.[39]

In another fascinating study, this time from the Cooper Clinic in Dallas, Texas, high levels of fitness and intensity decreased the risk of dementia, diabetes, stroke, and all-cause mortality.[40] They followed a group of 19,458 people from 1971 to 2009, and those in the highest quintile of fitness had a 36 percent lower risk of dementia.[41] That group included 26 percent women, so ladies, let's get with the exercise program.

Cure, for greater detail.) Signs that your hormones are in alignment include solid sleep, strong libido, consistent energy, and a lean body.

7. Intermittently fast, ideally twice per week.

8. Fill nutritional gaps such as missing zinc and B vitamins (including methylated folate).

9. Clean the air at your home and work. Test and remediate mold, a common cause of type 3 Alzheimer's disease and chronic inflammatory response syndrome (CIRS). Maintain and replace air filters monthly or as needed for optimal function.

10. Take supplements with neurotrophic and antioxidant effects.

11. Stimulate your brain to learn new things.[42]

12. Restore gut health by repairing leaky gut and eating nourishing foods for the microbiome. (Signs of leaky gut include chronic gas, bloating, constipation, diarrhea, headaches, fatigue, nutritional deficiencies, poor immune system, memory loss, and food intolerances.)

One of the best ways to preserve and improve your brain is to exercise regularly, probably because it upgrades your brain plasticity, neurogenesis, metabolism, and vascular function, which results in the release of growth factors like BDNF and improved memory and learning.[43] Physical exercise prevents neurodegeneration and keeps your hippocampus from shrinking and your cognition from declining, even if you are at genetic risk.[44]

Mind Makeovers for Weight Loss

Alzheimer's may be the scariest reason to take care of your brain and mind, but other mental-health issues affect aging.

It is time to address the ways that genetics can affect your behavior with food and, consequently, your weight, and to learn how to work around these tendencies. If there are a lot of people in your family who are over-weight or obese (or they work tirelessly to prevent it), listen up, because this is a matter of brain function too. (See appendix for full names.)

- ANKK1/DRD2 increases food desire, overeating, and addictive behaviors. Find pleasure in non-food items, like getting a regular massage, smelling fresh flowers in the house, taking hot baths, practicing yoga or meditation, and drinking green tea (which raises dopamine levels in the brain). Keep a food journal to stay on track and not let the inner addict run the show.

- MC4R is a gene variant that leads to over-snacking. Once you start, it can be very difficult to stop, like an avalanche. Those with the genetic variant are nodding along right now. The gene is expressed in the brain's hunger center and is linked to obesity. This is the way I handle it: I eat my first meal most days around 7:30, followed by lunch four to six hours later, followed by dinner four to six hours after that. Then I fast three hours before I go to bed (I stop eating by 7:00 P.M.). So I eat three times per day, no snacks. (When intermittently fasting, I eat my first meal around noon, and another four to six hours later.)

- The Fatso gene (FTO) as discussed leaves you having a hard time feeling full. For me, it's as if someone cut the wires between my stomach and my brain, so I don't know when to stop eating. Intuitive eating just doesn't work. I have found that the best work-around is to weigh food and commit to the type and quantities in advance, usually the night before. I know, I know, but it can keep you from overeating. The surprising part is that for me and some of my patients, this process provides more peace. No bargaining or wondering if I'm hungry.

- SLCA2 is a gene variant that gives some of my patients a sweet tooth. These women have an increased likelihood of eating more sweets that tend to go straight to their waistlines and cause brain fog. Our work-around is to include sufficient fruits in their meal plans.[45]

Integration of Mind and Body to Improve Healthspan

The other key to improving your thinking and healthspan is activating the COMT gene. Your goal is to turn off the worrier gene and turn on the warrior gene, start noticing your thoughts, identifying the negative ones on a regular basis. Becoming an objective witness to your thoughts is imperative. The reason why is that self-talk is often negative and self-defeating and can easily become part of a pattern of neuroticism or negative emotionality.[46] It's like the automatic-pilot setting; it just happens.

Self-talk can lead to either neuroticism or healing. That is, it can be harmful or healthful. Neuroticism refers to the tendency to respond with negative emotions to threat, loss, or frustration. For people who are neurotic, that response is frequent and out of proportion to the circumstances. Not surprisingly, neurotic traits increase your risk of poor health- and life span because they can create anxiety, depression, insomnia, and heart disease.[47] I used to joke that I could think my way into hormonal imbalance. Insomnia is particularly correlated with internalization, perfectionism, and an anxious or depressive coping style.[48]

The mind works best as your servant, not your master. Negative thoughts feel familiar, like a security blanket, so we become accustomed to them—but that doesn't make them good or healthy. It's like being in love with a terrorist . . . eventually things will go bad.

You need insight in order to work with the emotional, psychological, and spiritual dimensions of your experience, and it starts by addressing the biological basis. When it comes to retraining the mind, very few people are nutritionally and hormonally balanced enough to benefit from psychotherapy or coaching. If you're deficient in omega-3 fatty acids, B vitamins, sex or thyroid hormones, it will be hard for you to focus on your mind and soul. Reason and revelation together help us understand

Become a Psychoanalyst at Seventy-Four?

Rita Sussman, Ph.D., is a child psychologist and now psychoanalyst who revels in people and stories. She works part-time providing psychotherapy and psychoanalysis to adults and children and focuses on the interplay between emotional development and learning. Dr. Sussman was raising her children, including my close friend Jo, when she went back to school to complete a Ph.D. in 1979. Her dissertation was on curiosity and exploration, "twin pursuits that have been defining currents" in her life and likewise predict longevity.[49] At Jo's home near mine in Berkeley, Dr. Sussman is a frequent visitor who loves to spend time with her grandkids. She completes her patient care on Thursday afternoon and flies to Oakland to hang out for the weekend. While Jo and I freak out about the usual clamor of busy lives as working moms—unruly kids, overwhelming work/family demands, irritation at the impossible volume of work to be done—Dr. Sussman listens quietly and always offers a kind word, a calm presence, and wisdom that reflects her rich inner life.

Last year, at age seventy-four, Dr. Sussman completed the ten-year process of becoming a psychoanalyst. "Dealing with my dreams, the timelessness of the unconscious, and working through so many points of fragmentation in my life, gradually enabled an emerging sense of robustness and vitality. Developing a relationship with my second analyst allowed me to take pleasure in my accomplishments—indeed trumpet them—and allowed me a wonderful opportunity to believe in the development of my own self as a separate, autonomous being, connected to others, but not needing to be controlled by them or how they felt about me."

Dr. Sussman has no plans to retire. Her commitment to lifelong learning, curiosity, and being of service to her children, grandchildren, and patients keeps her mind sharp and young: "I'm more willing to *accept* where I am and recognize what I haven't done, and recognize that now I am living actually *who I am,* not who I want to be." Similarly, I hope you can connect to the timelessness of your unconscious. I want you to find the new idea or passion that you can commit to as a means of keeping your mind young and agile.

ourselves. My advice: correct the imbalances first with this week's protocol, then address your negative thoughts with the meditation approaches covered in chapter 10. It turns out that you have more influence than you may realize over your thoughts and how they affect your healthspan.

Protocol for Week 7: Think

It's time to make your brain and mind your partners. Most of us don't know which APOE variant we have, so the key is to act as if you do have APOE4 and commit to the lifestyle interventions that will turn it off. It will improve your inputs and may de-age your brain while boosting your healthspan.

Guess what—you're already doing many of the basics from earlier in the Younger protocol that improve your brain health and function.

- Keep sleeping seven to eight and a half hours per night, and nap if you get less. Sleep prods the glymphatic system to wash away the beta-amyloid plaques and remove other waste.

- Continue to exercise four to six days per week for at least thirty to forty minutes at a moderate pace, working in several high-intensity intervals, in order to raise BDNF, prompt neurogenesis, and lower blood sugar.[50]

- Are you still flossing twice per day and brushing three times, with an electric toothbrush? Yes, it prevents cognitive decline (see chapter 5).

- Keep improving mitochondrial function by eating more vegetables and repleting your antioxidant stores, avoiding refined carbohydrates, and intermittently fasting to activate longevity genes.

- Keep taking vitamin D_3 and fish oil each day.

Next we're going to add some of the most important actions to enhance the state of your brain.

Basic Rituals

- Food first. Dietary choices directly affect brain function and help stave off the brain enemies of diabetes and Alzheimer's.

  - Cashews are rich in magnesium (which may help sleep and prevent muscle tightness) and zinc (which may help memory). Add a handful as an appetizer daily or throw ten into a green shake as a meal replacement.

  - Quinoa is an ancient seed that is another great source of zinc and folate, which may help prevent dementia. Include it in at least two meals this week.

  - Turmeric root is anti-inflammatory; chop and add to quinoa in at least two meals this week. If you can't find turmeric root, use turmeric powder sprinkled on food.

  - Coconut oil[51]—a single serving of medium-chain triglycerides (the primary fat found in coconut oil) improves cognitive function in patients with memory problems.[52] A small randomized trial showed improvement in cognitive impairment at a dose of 56 grams per day (about a quarter cup). Use one tablespoon per serving at least twice this week, perhaps when you make quinoa with turmeric.

  - Eat probiotic-rich foods such as kefir, yogurt, kimchi, kombucha, and miso in at least two meals this week.

- Skip the restaurant and takeout this week. Eating eleven to fourteen of your meals at home reduces your risk of diabetes and obesity, which clobber the brain![53] The reason is that you have more control over ingredients and quantities; hopefully you don't use industrial oils while cooking, and you're more likely to stop when you're full.

- Create new neural pathways by using your nondominant hand for activities such as eating, writing, and brushing your teeth. The more new neural pathways you create, the more remembering you do in the balance of remembering and forgetting.

- Make bone broth. It helps to seal the leaks in your gut so that you can make the proper amount of happy-brain chemicals. I use it to make simple soups: I'll bring one quart of chicken bone broth with one head of cauliflower, chopped, to a boil and simmer twenty minutes, then puree with watercress in my blender, add a little sea salt and ground pepper, and garnish with ancho chili powder. See the appendix for recipes.

- Haven't tried the oil pulling from chapter 5? Here's your second chance, because it's so effective. Do it at least once this week to reset the microbiome in your mouth. Extra credit for doing it daily, then revel in well-deserved pride when your dental hygienist asks what you've been up to because your gingivitis is gone and there's very little plaque on your teeth.

- Smell something essential. Essential oils are proven to alter the biochemistry of your nervous system in powerful ways. Some oils stimulate while others calm you. If you're feeling low and unmotivated, try inhaling a stimulating scent such as grapefruit, black pepper, or fennel, which double your sympathetic nervous system activity.[54] If you need to calm down, try lavender or rose oil.[55] If it's easier, buy a bunch of fragrant roses and inhale the aroma all week long; smelling roses has been shown to improve physiological and psychological relaxation.[56]

Supplements

Take enough Methylcobalamin to keep your serum level at about 500 pg/mL or above—typically a dose of approximately 1 mg/day for most people, including Patient Zero. If you live in Japan or Europe, that's the recommended serum level (500–550 pg/mL), but here in the United States, the cutoff for deficiency is 200 pg/mL.[57] No wonder we have higher rates of dementia. There's a growing problem of marginal B_{12} serum levels, with 39 percent of people aged 26 and 83 in the low-normal range where people experience fatigue, memory problems, and other neurological symptoms.[58] Rates of low-normal levels are increased among vegetarians and vegans.[59] You need it for DNA synthesis,[60] to make red blood cells, and for bone strength. Consult with a knowledgeable clinician because many more nuances exist in interpreting your levels and how vitamin B_{12} is trafficked in your body.

Advanced Projects

Remember, we're trying to stop the leaks from thirty-six holes in the roof. The more you take on this week, the more likely you'll prevent or reverse the big "A."

- Try a new form of exercise this week. Use the Younger protocol to continue breaking out of your old patterns with fitness—it'll create new neural pathways. Ideas: start jumping rope; go to a spin, barre, or Pilates mat class; sign up with a trainer and learn high-intensity interval training with weights. Quick reminder: higher intensity exercise reduces your risk of Alzheimer's disease.

- Take more supplements.

  - Citicoline (CDP choline), a dietary supplement used commonly in Europe, may help your brain if you're lower than normal in your mental functioning.[61] As little as a single dose can improve processing speed, working memory, verbal learning, and executive function in healthy people with low-baseline cognitive functioning. Despite that, it has no effects in medium-baseline performers and it impairs high-baseline performers. Until we know more, only take it if you have cognitive impairment such as from dementia, acute ischemic stroke, or acute concussion; dose 500 to 1,000 mg. (In acute ischemic stroke, 2,000 mg.[62])

- CoQ10 will feed your mitochondria. Take 100 to 200 mg per day.

• Talk to a functional medicine clinician, hormone expert, or gynecologist about whether you're a good candidate for getting your hormones in balance. In particular, you want your estradiol, progesterone, and testosterone in the healthy range. Under proper care, you may improve memory, libido, and energy while staving off Alzheimer's disease.

• Pick a new cognitively stimulating activity from the following list and perform the task for around forty-five minutes at least twice this week.

- Learn a new language.

- Bake or cook (see appendix for brain-enhancing recipes).

- Take up a new musical instrument.

- Complete a crossword puzzle.

• Test your methylation capacity—your ability to methylate genes to turn them off. You may be methylating too much or too little.

• Assess your microbiome.

• Try NeuroRacer, a therapeutic video game created by Professor Adam Gazzaley of the University of California at San Francisco. Video games are no longer just for teens; Gazzaley designed NeuroRacer to combat age-related mental decline by using neurofeedback and TES (transcranial electrical stimulation) to boost the brain. The game requires that you steer a virtual car while performing other tasks. After playing for twelve hours (not consecutively!), older folks improved so strikingly that they were edging out twenty-year-old novices. The game specifically aids working memory and attention span, and improved skills that were transferable to the real world.[63] (See Resources for details.)

- Join a community or build one around your interests. I have a yoga community, a school community around my daughters' schools, a functional medical community, and, very excitingly, a community growing around the Younger protocol. As you may have read in the science section, community is a powerful lever for remembering and keeping your brain and mind healthy as you age.

- Try nutritional ketosis with high fat (70 percent), moderate protein (20 percent), and low carbohydrates (10 percent). In my opinion, nutritional ketosis works better for men than women and may worsen thyroid and adrenal function. I suggest trying nutritional ketosis in collaboration with a knowledgeable health practitioner who can track your results and help determine if it's a good fit for you.

What to Take After a Brain Injury

When I experienced a traumatic brain injury in 2015, I had a functional neurologist named Dr. Jay Lombard on speed dial. Here are the immediate treatments that he recommended:

- CDP choline, a supplement that acts as a neurotroph (makes the brain grow)

- Allopregnanolone, a hormonal metabolite of progesterone (You have to take it sublingually because it won't get into your body if you swallow it.)

- Zofran, if you're like me and experience nausea as a sign of brain inflammation

- Depakote, an antiseizure medication that is available by prescription only

Recap: Benefits of Week 7

Within one week, you may experience improved memory, focus, and mental acuity. Carried out over the long term, you can expect this week's changes to reduce your risk of accumulated oxidative stress and other bad pathways of neurodegeneration so that you can reverse or prevent cognitive decline and lower your chance of early death.

Bottom Line

Richard Dawkins says that genes created us, body and mind. It's true that they are the key to your past, present, and future. But they aren't the only factor. Once you're born and start making choices, it's your balance of inputs and outputs that create a brain and mind that are fit for a long healthspan. Yes, your genes are set for life, but you can change how the environment interacts with your genes. Your physical healthspan will never be at its best if your brain, mind, and nervous system aren't at their best. When you clean out your head space, you'll be able to create who you want to be in the moment and stay mentally sharp for years, allowing you to consistently choose the best actions that keep the aging process in check.

CHAPTER 12

Integrate

Challenges in medicine are moving from "Treat
the symptoms after the house is on fire" to
"Can we preserve the house intact?"

—Elizabeth Blackburn

I mentioned famine genes previously, but the true story of famine and its long-term effects is now known. Something scientifically stunning came out of a terrible period of suffering. I'm not talking about the mild hunger of my in utero experience, but a profound lack of calories experienced by four and a half million people during a Dutch famine known as the Hongerwinter ("Hunger winter").

In the German-occupied Western Netherlands, a bitterly cold period began in November 1944. German blockades of transports led to a catastrophic drop in the availability of food. People barely survived on about 30 percent of their normal calories. They were so desperate that they dug up tulip bulbs to eat. In Amsterdam, rations were sometimes less than four hundred calories in a day.

By the time the Allies liberated the region, in May 1945, twenty thousand people had died of starvation. When I hear about something so painful, I try to find the silver lining if I can. In the case of the Dutch famine, we have a window into famine epigenetics because it was a specific period that occurred in a location with meticulous health records. Women who were pregnant before, during, or immediately after the famine, as well as their male partners, were carefully tracked—now for more than seventy years—as were the 2,414 famine babies.

The scientific studies reveal that the babies born to those women of the Dutch famine had a normal birth weight depending on when in gestation they were exposed, but that their offspring (that is, the Dutch famine grandbabies) had more *neonatal adiposity* (they were fatter babies) and poorer health later in life. The parents' epigenetic changes influenced the genes of their children and grandchildren,[1] similar to what was found by Professor Rachel Yehuda's soul wounds of the Holocaust and 9/11 (see chapter 10).

As if soul wounds from a humanitarian disaster weren't enough, experts believe the exposure to famine accelerated the aging of the brain as the babies got to middle age.[2] In three hundred adults who survived the famine, the results of attention and advancing-age tests were worse sixty years after the famine compared with controls. Men but not women showed weaker hand grip, a sign of accelerated aging and the muscle factor, and more physical frailty.[3]

It turns out that famine exposure when a woman is one to ten weeks pregnant is the most dangerous time and associated with maximal DNA methylation.[4] As you may recall from chapter 3, methylation is like a sticky note providing instructions to DNA to do something different, like store fat or develop blood-sugar problems.

Another study looked at the undernourished fathers of the famine and found that the dads passed on a tendency toward obesity, suggesting that either the mother's or the father's epigenetics could leave a sticky note on a child's DNA.[5] Overall, the malnourished mothers gave birth to a generation that had more obesity, decreased physical and mental health, and earlier deaths than children who were not in utero during the famine.[6]

Remember Audrey Hepburn? She was an actress and humanitarian who won an Academy Award for the film *Roman Holiday*. She was born in

1929 in Belgium but grew up in England and the Netherlands, where she studied ballet. Hepburn was a survivor of the Hongerwinter and worked as a courier for the Dutch resistance. After living through the famine and despite being a fashion and film icon, she experienced lifelong poor health and died prematurely of cancer in 1993 at age sixty-four.

That's the power of epigenetics. However, just as genetics isn't a life sentence, neither is epigenetics. For instance, the timekeeping telomeres of the Dutch famine babies were normal, not shortened, as might have been expected.[7] Somehow their bodies fought off some of the potential damage. The human spirit is remarkably adaptable and resilient. That's why leveraging epigenetics is so important for your healthspan.

Connecting the Dots to Outfox Aging

The trouble for most of us is that we aren't making the connections between daily lifestyle choices and how our genes are expressed, so we don't even know what we can leverage. We don't always track what we're doing with what's happening in our bodies. We don't connect the excessive drinking in our forties and fifties to how it may raise the level of bad estrogens and risk of breast cancer. We don't understand that stress causes high cortisol and, eventually, depression, insomnia, and high blood pressure. We don't associate reading a good digital book until late into the night with the disruption of circadian rhythm and sleep and, again, the greater risk of breast cancer. We often think these "innocent" actions are just part of living fully and enjoying it.

No wonder there's a lack of understanding and even a bit of denial about how we are damaging our epigenetics and the way our DNA is expressed. Regretfully, these choices add up to increase your health problems, accelerate aging, and may set you up for a scary future diagnosis. Even after a serious diagnosis, few women get a prescription that addresses the root cause or the lifestyle actions that got them there.

We all age and die, but the difference in the rate of aging and quality of life has to do with your choices—the epigenetic changes within your power. My friend Sandy is an example of putting that power into play.

Should We All Move to Marin County?

Sandy and I emerge from the vinyasa yoga studio after ninety minutes, drenched in sweat. We're having our favorite playdate: hot power yoga followed by a cup of tea and a lively discussion about health.

Looking at Sandy, you might suspect she is my age: she is glamorous with long and thick blond hair, and she appears to be at least a decade younger than her chronological age, which is sixty. Her healthspan score is a stellar 82. She lives in Marin County. According to the Institute for Health Metrics and Evaluation at the University of Washington, Marin has some of the best longevity in the nation: men living in Marin County have the highest life expectancy of all men in the United States (81.6 years), and Marin women have the second-highest life expectancy of all U.S. women (85.1), based on estimates between 1989 and 2009.[8] (Wondering where women live the longest? Collier County, Florida: 85.8 years.) More people exercise in Marin County than anywhere else in the country. They eat better, smoke less, and get rewarded for it with a long healthspan. Sandy is one of these people. She has always exercised regularly, and vinyasa yoga is her go-to form of movement for mind/body alignment.

Family history was a potent motivator for Sandy to alter her course. Despite genetic tendencies for hormonal cancer, including breast cancer in her mother at age forty-two and prostate cancer in her father, Sandy is spectacularly healthy and takes no medications. In fact, she strongly believes that there's no pill that fixes health problems. Sandy's mother lived until ninety but faced a long battle with neurological degeneration. She lost her ability to walk and was confined to a wheelchair. She experienced chronic pain and had to go to rehab to get off pain pills. She even lost control of her facial muscles. During her forties, Sandy observed her mom's decline and determined that it wouldn't become her fate.

Sandy's mother survived breast cancer, but she was not an exerciser. She grew up in New Jersey on a diet of meat and potatoes, along with lots of processed food. Her family of origin believed that health issues were inevitable as you age, all part of getting older. Sandy disagrees. She believes that what you put in your mouth matters and that eating is more than just filling up. Similarly, Sandy came to exercise on her own.

In her forties, Sandy experienced a wake-up call. She didn't want to age like her mother. At the time, she regularly dined at fine restaurants and ate whatever she wanted. She didn't eat many vegetables. She went to spin classes at her gym but didn't practice yoga or meditation. She had young kids and routinely woke up with a start at three in the morning. Her doctor measured her telomeres and found they were short. Then Sandy's memory started to decline, and she began to think more about aging and what she wanted it to look like: "I did not like what I saw my mother experiencing and believed that I was not destined to go down the same road. That's when I started seeing an antiaging doctor, looking at blood work, and understanding the need to have my body in balance, eat the right foods, and get the right amount of sleep."

Sandy changed the course of her own aging process. She got religious about eating more vegetables, seasonal and organic. She attended medical and nutritional conferences and became passionate about teaching healthy eating and cooking. She got her hormones back into balance with bioidentical hormone therapy. She took supplements based on her lab work. Her memory improved; her weight dropped. She felt better in her fifties than she ever did in her forties.

In fact, Sandy now has the zeal for food intimacy that I found in Betty Fussell. "I believe food is a critical part of this process. I have cut out all processed foods and removed toxins out of my diet. I eat mostly vegetables, fruits, nuts and seeds, and small amounts of animal protein (organic, non-GMO, grass-fed meats, wild fish). I use lots of fresh herbs. I cook at home so that I know the quality of the food I am eating. When eating out, I eat as many vegetables as I can find on the menu. I keep my weight level, and my energy is good all day. I avoid processed sugar. And when I want a sweet, I eat dark chocolate, and I don't eat a lot of it. I see many women my age who have gained that extra ten to twenty pounds around the midsection. They look puffy—it's inflammation that the wrong foods can cause. I stay away from gluten and don't eat much dairy. But my philosophy is that if I really want to eat something, I will. I am not in a food straitjacket."

What's Sandy's greatest fear? Cognitive decline. She is well aware of how to prevent it and prioritizes getting eight solid hours of sleep every night. She intermittently fasts overnight for twelve to fourteen hours. She takes vitamins regularly, keeping antioxidants up and inflammation levels down. Her hormones are in balance; she uses transdermal (applied to the skin) bioidentical estrogen and testosterone, together with oral progesterone (as a pill) to prevent the buildup of her uterine lining (which will prevent endometrial cancer). Her current supplements include the following:

- Multivitamin and minerals

- Vitamin B complex

- Vitamin D_3

- Vitamin E (not everyone needs this; it depends on your genes)

- Indole-3-carbinol (similar to the di-indole methane [DIM] mentioned in chapter 3)

- Trimethylglycine to help methylate away (inactivate) bad estrogens

- Alpha-lipoic acid, a powerful antioxidant

- Fish oil

- Borage oil

- Ubiquinol to replenish CoQ10 stores

- Phosphatidylserine to ease stress

- L-methionine to help make more glutathione and detox via the liver

- Probiotics to support the microbiome

Sandy turns off the genes for breast cancer and turns on the genes for longevity. I'm impressed with how Sandy follows the lifestyle changes needed to lose weight and thereby reduce her risk of breast cancer and early death. So far (fingers crossed), Sandy is preserving the house intact, to paraphrase Professor Elizabeth Blackburn, the Nobel laureate quoted at the beginning of the chapter. Now it's time to make your seven-week protocol into a plan for life.

Why It Matters

The main question of this book is, How can we live in a way that activates the genes for the best healthspan? By now you know that lifestyle is more important than genetics when it comes to your healthspan.

In the United States, more people are living to age one hundred. We call them centenarians. Their numbers have increased 66 percent since 1980; they are the fastest-growing demographic in the United States. Be one of them.

Here is my best advice for how to maintain your progress with boosting your healthspan:

- Follow the seven-week protocol in *Younger*. Lengthen healthspan through the way you eat, drink, sleep, move, release, expose, supplement, de-stress, and think.

- Retake the Healthspan Quiz after you complete the seven-week Younger protocol. Take it again after four to six months to see if you need to repeat the protocol (if your score declines). Aim to improve your score every year.

- Keep up the behavioral changes. Download an app, like Lose It or MyFitnessPal (available for free) to keep track of your weight, body fat, activity, and food all in one place. Or try the app Fitness Builders to keep track of your workouts.

- Eat at home. Avoid restaurants; most of them use industrial seed oils that provoke more inflammation in your body.

- Brush with an electric toothbrush and floss twice a day.

- Enlist a buddy to stay accountable.

- Remain vigilant against the environmental agents that may hurt your genes: pollution, mold, ozone, pesticides, skin care, cleaning products, sitting more than three hours per day.

- Meditate every day.

- Keep taking regular saunas. Aim for four times per week.

- Soothe your genes with regular stretching and, as your budget allows, someone who can help you release your areas of recurrent tightness.

- Invest in a stand-up or treadmill desk.

If most of the steps in the protocol are new to you, consider running your protocol first with just the basic rituals, then repeat in four to six months but add one or more of the advanced projects. Healthspan is a complex equation with many variables. By following the basic protocol (the steps get easier as they become habits), you'll address the most important ones and develop a Younger code for how to age gracefully. (See a functional medicine health professional if you're not seeing the results you hoped for or you want to track your progress more closely.)

Your Daily Routine

The next page shows Sandy's daily routine to show you another way the Younger protocol plays out. Modify to carry forward your best healthspan.

Time to Change Our Attitude Toward Aging

Even with as much change as you've been implementing over the last seven weeks, I want to address one more crucial mental element. Maybe it takes turning fifty to notice all the negative stereotypes about aging, but we're saturated by them. Advertisements, health books, TV shows, and daily conversations are rife with bad connotations: *old* is frail, demented,

A TYPICAL DAY IN THE YOUNGER PROTOCOL: SANDY

7:00 A.M.	Wake up and brush teeth with electric toothbrush
	Take thyroid medication (Nature-Throid)
7:20	Prepare either lemon juice and warm water, or an almond milk latte (with homemade almond milk) or ginger juice (homemade with lemon in warm water)
8:00	Eat berries plus kefir chia pudding or nuts
9:00	Hot yoga and drink water
10:30	Apply estrogen and testosterone after showering
	Take B12
12:30 P.M.	Lunch is either leftovers, or a salad, or soup and salad
1–5:30	Work
5:30	Cook dinner (mostly vegetables, some protein, usually fish or chicken)
6:30/7:00	Eat dinner
9:00	Take an Epsom salts bath and relax
	Take progesterone and supplements (she takes them at night because it's easier to remember and easier on her stomach)
	Brush, floss, and cleanse. (Sandy uses natural toothpaste and an organic cleanser, and generally keeps all products on her skin clean. She uses sesame oil on her body, a face cream containing vitamins C and E, and all-natural makeup.)
11:00	Go to bed

helpless, incompetent, undesirable, unattractive. Sadly, these messages can become self-fulfilling prophecies for the old and young alike.

My mother said something very astute the other day. She explained that in the current era of the high-glam, high-disclosure content online (think of the splashy celebrity families on TV and social media) and anorexic teenagers in fashion magazines, older people feel dismissed, invisible, marginalized, and alienated. Mom asked me, "When was the last time you saw a sixty-year-old on the cover of *Vogue*?" I thought of Hillary Clinton, but that was certainly an exception for a glamour magazine. Mom made an important point: while older folks are not represented positively in media, they're the ones with the buying power. "We're dismissed by doctors and the mainstream press. It's important to change this. We care, we won't be silenced, and we matter." Mom's well-reasoned rant resonated with me, and it prompted me to look further.

Science backs up her observation: our current inundation with negative age stereotypes makes older people feel worse about themselves and predict worse physical function, including memory and cognition.[9] In short, bad media stereotypes about aging actually make older people even more frail and forgetful. Women are more affected than men by negative imagery toward aging. Look at *Esquire* covers in the past few years: Clint Eastwood, Robert De Niro, Donald Trump, Michael Keaton, Liam Neeson—all over sixty. Aging men are seen as sexy and wise, role models for younger men and all of us. Women get the short end of the aging stick. That I can't abide.

Can we undo the damage of the negativity and how it affects the exposome? Yes, we can. I discovered a fascinating study that found implicit positive age stereotypes improve physical functioning more than exercise does![10] How's that? The intrepid group that did the study had previously found that showing positive age stereotypes countered elders' beliefs about their physical function.[11] Showing older people positive images of themselves also fed something important: their preference for more positive images.[12]

Since their first study, these researchers have also discovered that showing aging in a positive light bypassed older people's internalized negative stereotypes, improved their own stereotypes of aging, and upgraded their self-perception and their physical function. The authors put it best: "These findings suggest the intervention served, in effect, as an implicit fitness center." Wow! Positive images literally improved their strength, gait, and balance! That's the kind of positive exposure I can totally get behind.

Essentially, we must change the stereotype and create the new and positive "gym." Aging can be beautiful, healthy, and strong. Let's see more of Betty Fussell and Ida Keeling featured again in *Vogue*, defying my own stereotypes and attitudes about aging. I'd love more media images of Madonna, fifty-seven at the time of this writing, and her peers in advertising campaigns, such as we see for Versace and Louis Vuitton. We need more photos of iconic author Joan Didion, currently eighty-one, considered by some to be the most discussed fashion model of 2015[13] (in an advertisement for the French label Céline). As one journalist put it, "The fashion house's use of the writer as the star of its new ad equates brains with beauty. Hurrah."[14]

People who have an optimistic self-perception of aging live eight years longer.[15] Not to mention the fact that their minds and hearts will benefit from embracing the wisdom and clarity of added years.

Bottom Line

As we have learned, it's not genes alone that carry you into healthy old age. Although you can't change your genes, you can change how the environment interacts with your genes to determine who you are in the moment and each new day. That's because epigenetics is mutable and reversible. But you need an effective plan, which may shift as you get older and as researchers make more discoveries about gene-environment interaction. Your forties may be when you notice more gray hairs and consider covering them with hair dye or examine your face and wonder about injecting fillers, but the truth is that prioritizing sleep, exercising, flossing, finding meaning and fostering connections, staying curious, consuming more antioxidants, and bathing in a sauna are far better choices as you age.

Your needs and sensitivities change with age. What works for you at forty may not be what works for you at fifty or sixty or seventy. So I hope you've found enough options within the Younger protocol to personalize it for you. What's important is the sentiment expressed by Dr. Rita Sussman: "If one is fortunate, aging can continue to provide exceptional pleasures, new engagements, and the continuing use of one's mind and heart to help others and deeply experience joy in one's life." *Amen.*

Overall, my greatest wish is for you to make the principles foundational habits. I've designed the protocol to be done two or three times per year, the second and subsequent times with the more advanced projects added.

After a year of living by the Younger protocol, you'll reach a new state of homeostasis, or biological equilibrium, with the forces that age you versus the forces that slow down aging. You'll need to readjust every year to keep the biology of aging on your side. That's the challenge and the promise of functional medicine: you can adjust your environment and lifestyle in accordance with your ever-changing baseline.

To help you stay connected to the concept of healthspan and live it daily, repeat your healthspan quiz, track your results over time, and endeavor to improve your score each year.

As we close the book, I want to offer another wise nugget from Sandy in Marin. I asked her about fear and death, about her *why* for boosting healthspan. She replied: "I don't think about my death. I do think about the quality of my life and the amount of time that is left. My philosophy is to try to be in the moment now, to find a higher reason for being, and to live my passion."

Remember to connect with your *why* for lengthening your healthspan. Just as what works for you biologically now may be different than what worked for you biologically five years ago, your *why* may change over time. What's your *why*? Are you living it each day? Are you getting the self-care that you need, broadened now to include a long healthspan?

Keep your eyes on the prize—leverage epigenetics and the $^{90}/_{10}$ rule to work for you. Use the changes of *Younger* to clean up the neighborhood for your genes so you are able to live joyfully, free from disease, in hormonal balance, and brimming with youthful vigor.

Recipes

Food is information for your DNA. It's also meant to be delicious and nourishing. These recipes are designed to lengthen your healthspan and are easy for busy people to prepare. Each one has been carefully tested and tweaked in my kitchen.

THE YOUNGER PROTOCOL SHAKES

Dr. Sara's Younger Breakfast Shake

1 cup iced green tea or matcha (page 273), choose decaffeinated green tea if
  necessary
½ cup unsweetened almond or coconut milk
2 scoops Dr. Sara's Reset360 All-in-One Shake, vanilla or berry
1 tablespoon medium-chain triglyceride (MCT) oil
2 tablespoons hemp seeds
1 tablespoon freshly ground flaxseeds
1 scoop Dr. Sara's Super Greens powder
1 to 2 teaspoons maca powder
6 to 8 ice cubes

Whip all ingredients in a high-powered blender (e.g., Blendtec, Vitamix, or Nutri-
Bullet) to desired consistency.

Dr. Sara's Think Shake

1 cup unsweetened coconut milk with cream
1 tablespoon MCT oil
2 scoops Dr. Sara's Reset360 All-in-One Shake, vanilla or chocolate
½ avocado
5 Brazil nuts
1 tablespoon hemp seeds
1 cup broccoli sprouts (instructions for growing your own on page 276)
1 to 2 cups frozen spinach

Whip all ingredients in a Blendtec, Vitamix, or NutriBullet.

Dr. Sara's Cashew Cacao Shake

8 ounces full-fat organic unsweetened cashew milk
2 scoops Dr. Sara's Reset360 All-in-One Shake, chocolate
½ cup kale, spinach, and/or broccoli sprouts
5 ice cubes
1 tablespoon cacao nibs for topping (optional)

Whip in a high-powered blender. Top with cacao nibs, if desired.

OTHER YOUNGER PROTOCOL DRINKS

Coconut Milk Coffee Layered with Avocado

1 cup freshly brewed low-toxin coffee (such as Bulletproof)
2 tablespoons coconut cream
½ avocado, sliced

Whip coffee and coconut cream in a high-powered blender. Layer with slices of avocado in a glass or jar.

Inspired by Kopi Kopi in Greenwich Village, New York
http://thechalkboardmag.com/new-york-cofee-shop-kopi-kopi

Matcha

With matcha you're actually drinking the entire green tea leaf, not just the tea water; this is one of many reasons why matcha tea is much more nutrient-dense than standard green tea. Matcha tea is very high in antioxidants, amino acids, and chlorophyll, which is responsible for its distinct bright green color. L-theanine is the most prevalent amino acid; it increases serotonin, dopamine, and GABA and is known to have a calming effect on the mind and body (likely why traditionally monks sipped matcha tea). The caffeine content may promote focused energy without the jitters.

1 teaspoon matcha tea powder
½ cup hot water (not boiling)
½ cup unsweetened coconut milk
Few drops of stevia to taste (optional)

Add the matcha tea powder to a small amount of hot water in a matcha tea bowl or your favorite mug. Using a bamboo whisk (or small metal whisk), briskly blend in an up-and-down direction to make a thick, green paste. Then pour the remaining hot water and coconut milk into the paste and stir. If you require a sweetener such as stevia, add here. The matcha will dissolve quickly and easily. If you're using a milk frother, place on the top of the latte and turn on, allow to froth and foam until your desired texture.

You may also use almond milk or sprinkle with cinnamon. Drink hot, warm, or chilled!

Green Tea Frappuccino

½ cup coconut milk or other milk of choice
½ cup water
1 cup ice
2 teaspoons matcha green tea powder
1 to 2 teaspoons xylitol or other sweetener of choice
½ teaspoon pure vanilla extract or pure vanilla bean powder

Place all ingredients in a blender and blend until smooth.

Turmeric Latte

4 tablespoons raw cashews
4 tablespoons shredded unsweetened coconut
1 cup water
1 teaspoon coconut oil
½ teaspoon cinnamon
1 teaspoon turmeric
Pinch of clove
Pinch of coarse sea salt
Pinch of cinnamon

Blend the cashews, shredded coconut, and water till creamy. Strain through a nut-milk bag and discard the pulp (you now have cashew milk). Put the liquid back in the blender with the rest of the ingredients and give it a quick whiz. Transfer to a pot on the stove, bring to a boil (or heat gently until warm to the touch), remove from heat, and serve warm with a dusting of cinnamon.

Note: You may also like to try using coconut milk in place of the cashew milk.

Dr. Sara's Beauty Tonic

2 celery stalks
½ cucumber
2 cups kale
1-inch piece ginger root
½ cup parsley
¼ cup blueberries
½ avocado
Pinch of ground cinnamon
Pinch of matcha tea
1 tablespoon fresh lemon juice
1 tablespoon chia seeds
2 scoops Dr. Sara's Reset360 Vanilla Shake powder
Water and ice to taste

Wash vegetables thoroughly. Place all ingredients in a high-powered blender and blend until smooth. Serve immediately.

Sexy Sangria

Quench free radicals that accelerate aging with antioxidants.

  1 bottle red wine (preferably organic)
  3 oranges, thinly sliced
  1 teaspoon lemon zest (I prefer Meyer lemons)
  1 lemon, thinly sliced
  1 lime, thinly sliced
  ¼ cup pomegranate seeds
  ½ cup raspberries
  1 to 2 quarts sparkling water (optional)
  Sprigs of rosemary or crushed pink peppercorns, for garnish

Mix all ingredients together in large pitcher. Chill in fridge and add ice. Garnish with sprigs of rosemary and crushed pink peppercorns.

BIGGER PROJECTS

Broccoli Sprouts

Yields about 4 cups

> 2 tablespoons organic broccoli sprouting seeds
> Wide-mouthed quart jar with spouting lid
> Purified water

Place the seeds into a jar and cover them with a few inches of warm purified water. Let them soak overnight in a warm dark place. After about 8 to 10 hours, drain the water off. Rinse the seeds with fresh water 2 to 3 times a day for 4 to 5 days. Place the jar in a warm, dark place during this time period. Make sure to drain off all the water after each rinsing to prevent spoiling of the sprouts. It will likely take the seeds 2 to 3 days to split open and begin to sprout, so be patient. Once your sprouts are about an inch long and have defined yellow leaves, move your jar out into a place where it can be exposed to some sunlight. This will allow the sprouts to use the light and grow quickly. Be sure to keep rinsing, as the sprouts can dry out quickly in hot environments. You will recognize when the sprouts are ready because they will have darker green leaves and be about an inch or longer. Don't worry about eating them too early. As soon as they are green, they are ready to consume.

By Tom Malterre
Reprinted from https://wholelifenutrition.net/articles/recipes/how-make-broccoli-sprouts

Alkaline Broth with Collagen

Give yourself a facelift right out of your fridge.

1 to 2 cups of three of the following vegetables:
    Celery
    Fennel
    Green beans
    Zucchini
    Spinach
    Kale
    Sorrel
    Chard
    Carrots
    Onion
    Garlic
    Cabbage
Fresh or dry spices (e.g., cumin and turmeric)
1 tablespoon powdered collagen protein (Bulletproof or Great Lakes are great brands)

Place vegetables and spices in a large soup pot and cover with filtered water. Bring to a boil and simmer on low for forty-five minutes. Strain the vegetables and set aside for another use. Whisk in the powdered collagen protein.

Adding Fermented Flavor

Consider adding fermented food such as kimchi to any meal. Kimchi is a spicy Korean version of sauerkraut typically consisting of fermented cabbage, onions, garlic, and pepper. It's been shown to lower fasting blood sugar. Kimchi has high concentrations of vitamin C and carotene in addition to vitamins A, B_1, B_2, calcium, iron, and beneficial lactic acid bacteria. Fermented foods are great for digestion and repopulate your intestinal flora with beneficial bacteria essential for health.

Fish Bone Broth

In Chinese medicine, the adrenals are considered part of the kidney system. Bone broth detoxes and nourishes the kidneys. Fish stocks that use fish heads contain thyroid-strengthening properties.

3 quarts filtered water
2 pounds fish heads and bones (fish heads alone will suffice)*
¼ cup raw, organic apple cider vinegar
Himalayan or Celtic sea salt to taste

Place water and fish heads/bones in a 4-quart stockpot. Stir in vinegar while bringing the water to a gentle boil. As the water begins to boil, skim off any foam that rises to the surface. It is important to remove this foam, as it contains impurities and off flavors. Reduce heat to a simmer for at least 4 hours but no more than 24 hours. Cool and then strain into containers for refrigeration. Freeze what you will not use in one week. Add salt at the end, to taste.

*Do not use oily fish such as salmon for fish stock or you will stink up the whole house! Use only non-oily fish such as sole, turbot, rockfish, or, my favorite, snapper.

Chicken Bone Broth

1 chicken (chicken bones, chicken feet, neck)
2 small onions or shallots
1 head garlic
1 teaspoon peppercorns
1 or 2 bay leaves
2 tablespoons sea salt
2 tablespoons apple cider vinegar
4 quarts filtered water
1 bunch fresh organic herbs (e.g., tarragon)

Put all the ingredients except the fresh herbs into a large stockpot and let sit for an hour. Bring to a boil and get rid of any foam that rises to the top. Cook on a very low flame for 8 to 12 hours. Let cool. Separate the meat (if any) from the bones. Strain the broth. Wash the fresh organic herbs well. Warm up a serving of strained broth to desired temperature (do not boil). Add a large handful of herbs (for extra minerals and taste).

Collagen-Boosting Chicken Soup

Makes 6 servings as a meal

STOCK

1 whole free-range chicken (ideally pasture-raised)
4 quarts cold filtered water
2 tablespoons vinegar
2 large onions, coarsely chopped
3 carrots, peeled and coarsely chopped
4 celery stalks with leaves, coarsely chopped
4 peeled garlic cloves
2 well-washed leeks, coarsely chopped
3 parsnips, coarsely chopped
3 bay leaves
4 to 5 sprigs fresh thyme, or 2 teaspoons dried thyme
10 whole black peppercorns
1 bunch parsley

SOUP

2 quarts chicken bone broth
2 cups cooked chicken
2 onions, chopped
3 cups celery, chopped
3 carrots, peeled and sliced
1 cup green beans
3 cups fresh spinach
6 cloves garlic, minced
1 teaspoon dried thyme
2 teaspoons sea salt
$\frac{1}{2}$ teaspoon freshly ground black pepper

Place chicken into a large pot with water, vinegar, and all of the ingredients except the parsley. Let stand 30 minutes to 1 hour. Bring to a boil, and remove any foam that rises to the top. Reduce heat, cover, and simmer for 6 to 24 hours. The longer you cook the stock, the richer and more flavorful it will be. About 10 minutes before finishing the stock, add parsley. This will impart additional minerals to the broth. Remove the chicken, let cool, and remove the meat from the carcass. Reserve for soup.

Strain the stock and reserve in your refrigerator until the fat rises to the top and congeals. Skim off this fat and reserve the stock in covered containers in your refrigerator or freezer. For the soup, bring 2 quarts chicken stock to a boil and skim off any foam that may rise to the top. Add the meat, vegetables, and seasonings and cook until the vegetables are just tender, 5 to 10 minutes.

Taste and adjust seasonings. Don't forget you can sip broth like tea. This is especially lovely in the winter or if you don't feel well. Since broth is simultaneously energizing and calming, it can take the place of morning coffee, afternoon tea, or an evening nightcap. Carry some of your favorite bone broth in a Thermos and sip it throughout your day. You'll discover the real meaning of "comfort food"!

Beef Bone Broth

> 2 pounds (or more) femur bones from grass-fed cattle or bones from
>   a healthy source
> 2 chicken feet for extra gelatin (optional)
> 1 onion
> 2 carrots
> 2 stalks celery
> 2 tablespoons apple cider vinegar
> 2 cloves garlic
> 1 bunch parsley, 1 tablespoon or more sea salt, 1 teaspoon peppercorns,
>   additional herbs or spices to taste (optional)

If you are using raw bones, especially beef bones, it improves flavor to roast them in the oven first. I place them in a pan and roast for 30 minutes at 350°F. Then place the bones in a large 5-gallon stockpot. Add the chicken feet. Pour (filtered) water to cover the bones plus a few inches, so the bones are fully immersed, and add the vinegar. Let sit for 20 to 30 minutes in the cool water. The acid helps make the nutrients in the bones more bioavailable.

Roughly chop the vegetables (except the parsley and garlic, if using) and add to the pot. Add any salt, pepper, spices, or herbs, if using. Now bring the broth to a boil. Once it has reached a vigorous boil, reduce to a simmer and simmer until done.

During the first few hours of simmering, you'll need to remove the impurities that float to the surface. A frothy/foamy layer will form and can be easily scooped off with a big spoon. Throw this part away. I typically check it every 20 minutes for the first 2 hours to remove this. Grass-fed and healthy animals will produce much less of this than conventional animals.

During the last 30 minutes, add the garlic and parsley, if using.

Remove from heat and let cool slightly. Strain using a fine-mesh strainer to remove all the bits of bone and vegetable. When cool enough, store in a gallon-size glass jar in the fridge for up to 5 days, or freeze for later use.

SALADS

Torn Greens with Younger Ranch Dressing

My family loves this dressing over grilled romaine hearts or as a dip for cucumber slices. Tearing greens increases the nutrient density.

FOR SALAD BASE
2 to 8 cups torn romaine lettuce, kale, spinach, or other greens

MAYONNAISE
1 cup avocado oil, olive oil, or a mixture
1 egg yolk
1 tablespoon Dijon mustard
Juice of ½ lemon
½ teaspoon salt

YOUNGER RANCH DRESSING
1 cup dairy-free mayonnaise (see recipe below)
¼ cup coconut milk
1 teaspoon apple cider vinegar
½ teaspoon onion powder
½ teaspoon garlic powder
1 tablespoon fresh dill or 1 teaspoon dried dill
2 teaspoons dried parsley or 2 tablespoons minced fresh parsley
1 tablespoon dried chives or 3 tablespoons minced fresh chives
Salt and pepper to taste.

Mayonnaise

Place all ingredients in a narrow container or jar. I use the mixing cup my immersion blender came with, but a half-pint jar works well too. Place the head of the immersion blender at the bottom of the jar and turn the blender on. The bottom of the jar should quickly emulsify (you'll see it turn white and thick). Slowly move the immersion blender up toward the top of the jar as the mixture emulsifies. If any oil slips back down into the jar, simply move the head of the blender down to mix it in, then continue lifting the blender up toward the surface until all the oil is incorporated and the mixture is thick. This process takes only 1 to 2 minutes at most.

This mayonnaise will keep covered in the refrigerator up to 1 week. It's great for chicken or egg salad, on a sandwich, or in creamy salad dressings.

Younger Ranch dressing
Add the additional dressing ingredients to the mayonnaise. Stir to mix well. Add additional coconut milk as needed to thin the mixture (it will naturally thicken somewhat in the refrigerator). Taste and add salt and pepper as desired. Pour over torn greens and toss.

Ranch dressing will keep one week covered in the refrigerator.

Seaweed Salad

Seaweed is very high in essential minerals that help love up your thyroid naturally, including iodine, calcium, iron, copper, magnesium, manganese, molybdenum, phosphorus, potassium, selenium, vanadium, and zinc. Some seaweed salads that you can buy premade have questionable sugar or not-so-great oils and vinegars. Here's one that's cleaned up so you can eat like a mermaid.

SALAD
2 ounces dried wakame (or a seaweed mix)
1 small daikon radish, julienned
½ English cucumber, julienned

DRESSING
1 teaspoon sesame oil
Juice of half a lime or lemon
2 teaspoons fresh ginger juice
1 tablespoon tamari
4 tablespoons walnut or avocado oil
½ teaspoon stevia or to taste
Pinch of sea salt
Toasted sesame seeds (optional)
Crushed roasted nori (optional)
Diced avocado (optional)

Soak seaweed in cold water for about 5 minutes until it's rehydrated and no longer tough. Rinse and drain. If you have any large pieces, feel free to give them a rough chop. Mix together dressing ingredients in a small bowl. Combine the seaweed, cucumber, and daikon.

Toss seaweed mixture and dressing well and let sit for a few minutes for the dressing to be absorbed. Add additional toppings if desired and serve with chopsticks.

Hail Kale and Caesar Salad

"RAW PARMESAN"
½ cup macadamia or cashew nuts, not soaked
1 teaspoon nutritional yeast (or more, to taste)
1 pinch garlic powder (optional)

DRESSING
½ cup cashews, soaked for 2 or more hours
¼ cup hemp oil
¼ cup nutritional yeast
Juice of 2 lemons
1 garlic clove, crushed
½ teaspoon sea salt or pink Himalayan salt
⅔ cup filtered water

LETTUCE AND VEGETABLES
1 bunch lacinato kale
2 heads romaine lettuce
1 cup cherry tomatoes, halved

To make the "raw Parmesan," grate or process nuts in a food processor. Add remaining ingredients and process until combined.

To make the dressing, rinse and drain the cashews. Combine the remaining dressing ingredients and blend until smooth.

Destem the kale and then finely chop the leaves. Wash and dry in a salad spinner. Place into extra-large bowl. Tear the romaine into bite-size pieces. Rinse and then spin dry. Place into bowl along with kale. You should have roughly 2 to 3 cups chopped kale and 4 to 6 cups of torn romaine.

Add dressing to lettuce and toss until fully coated. Season with a pinch of salt and mix again. Dressing keeps for up to one day in the fridge.

Turn Back Time Salad

Makes 2 large servings, or 4 small servings

Eat enough of this salad and you may start getting carded!

KALE SALAD
1 head lacinato kale
¼ cup pumpkin seeds
1 green Granny Smith apple, thinly sliced

CREAMY DRESSING
2 tablespoons macadamia nut or sesame oil
Juice of 1 lemon
1 large, ripe avocado
1 tablespoon tahini (sesame paste)
1 tablespoon hemp seeds
1 teaspoon minced garlic
2 tablespoons water (or more if needed to thin out)
¼ cup fresh cilantro leaves
Pink Himalayan sea salt and freshly ground pepper

Prepare your kale by washing it, removing thick stems, and finely chopping into strips. Place into a large bowl and set aside. For the dressing, place oil, lemon juice, avocado, tahini, garlic, and hemp seeds in a food processor. Pulse until smooth and creamy, adding in pink Himalayan sea salt and pepper to taste and water for desired consistency. Pour dressing over bowl of chopped kale.

Using your hands, massage dressing into kale for about 2 to 3 minutes, until leaves are completely smooth and silky. Add pumpkin seeds, cilantro, and apple slices and thoroughly combine. Serve kale salad immediately or store in an airtight container in the refrigerator for 2 to 3 days.

A Note on Oils

I love that macadamia nut oil is similar to olive oil in that it doesn't require industrial solvents or complex processes to extract the beneficial fat. It is low in saturated fat, but rich in the all-star monounsaturated and polyunsaturated fatty acids, which is what makes it so heart-healthy. Macadamia nut oil also has a properly balanced one-to-one omega-3 to omega-6 ratio.

Endive, Fennel, and Pear Salad with Walnuts

I remember tasting my first endive salad at Alice Waters's iconic Chez Panisse in the 1990s. This salad looks beautiful on a plate and features a fantastic texture and flavors.

½ **cup olive oil**
2 **tablespoons lemon juice**
Sea salt to taste
1 **tablespoon finely chopped shallots**
2 **teaspoons fresh thyme or 1 teaspoon dried thyme**
2 **Bosc pears (peeling is optional)**
1 **medium bulb fennel, trimmed**
½ **pound endives, trimmed**
¼ **cup walnuts, toasted**
Handful of pomegranate seeds (optional)

In a large bowl, whisk together the olive oil, lemon juice, sea salt, shallots, thyme, and a pinch of salt. Let stand for 10 minutes to develop flavor.

Thinly slice pears; cut the fennel into quarters and slice as thinly as possible by hand or with a mandoline. Separate the endive leaves. Whisk the dressing and gently toss it with the pear, fennel, and endive. Serve on salad plates topped with walnuts and garnish with pomegranate seeds if using.

Salmon and Avocado Salad with Miso Dressing

This cortisol-resetting-power salad will make you feel like a glowing goddess.

Serves 4

Olive oil
4 6-ounce salmon fillets (or steelhead trout, which tastes similar)
1 to 2 lemons
6 cups torn romaine lettuce
1 avocado, diced
¾ cup sliced cucumber
½ red bell pepper, thinly sliced
¼ cup walnuts, toasted

MISO DRESSING
2 teaspoons fresh lime juice
2 teaspoons white miso
2 teaspoons water
¼ teaspoon ground pepper
3 tablespoons extra-virgin olive oil

Preheat the broiler. Place the rack about 6 inches from the broiler. Line a baking sheet with foil and lightly coat the foil with olive oil.

Arrange the salmon fillets on the prepared sheet, skin-side down, rub with lemon juice and olive oil, and season with sea salt. Cook until the salmon is just cooked through, 7 to 10 minutes (depending on the thickness of the salmon). Remove the skin from each fillet. Chop into generous bite-size pieces.

While salmon is cooking prepare dressing: In a small bowl, whisk together the lime juice, miso, water, and pepper. While whisking, slowly pour in the olive oil.

Combine and toss lettuce, avocado, salmon, cucumber, and red bell pepper in large bowl. Divide between four plates. Drizzle one tablespoon of the miso dressing over each salad. Top with walnuts and serve.

MAIN DISHES

Egg Avocado Bake

Large avocados
Eggs (1 egg per avocado half, or 2 per whole avocado)
Hot sauce (optional)
Salt and pepper to taste
Cilantro, scallions, and hot chilies for optional toppings

Halve an avocado and remove the pit. Scoop out some of the flesh (about 1 tablespoon) so you have a hole big enough for your egg. Repeat with remaining avocados.

Put the avocados in a small baking dish. You want them to be sort of snug, so they don't tip over. It helps to nestle the avocados in pie weights, dried beans, or coarse salt to keep them standing up straight. Pro tip: Put a couple drops of hot sauce or whatever condiment you like in the hole before you add the egg. Crack one egg at a time into a small ramekin or glass. Slide an egg carefully into the hole of each avocado. Season with salt and pepper and drizzle any desired condiments over the avocado halves. I like to use a little pesto or chimichurri here. Bake at 450°F for 10 to 12 minutes, or until your egg whites are set and the yolks are still a little runny.

Douse with some greens or toppings (cilantro, scallions, and hot chilies are all delicious options)!

Buddha "Fettuccine" Alfredo

"FETTUCCINE"

2 extra-large turnips, spiralized

Shredded carrots (eyeball to serve you adequately; I like 1 cup)

Shredded, destemmed lacinato kale (eyeball to serve you adequately;
  I like 2 cups)

BRAZIL NUT SAUCE

6 tablespoons Brazil nut butter (½ cup Brazil nuts can also be substituted)

6 tablespoons water

2 tablespoons apple cider vinegar

2 tablespoons tamari

Sea salt to taste

Blend all the ingredients in a Blendtec, Vitamix, or NutriBullet on high. Pour about ¼ cup sauce over the vegetables, adding more if you need it to evenly coat them.

Plant-Based Fajitas

Makes 2¹/₂ cups

LENTIL-WALNUT MEAT
1 cup uncooked French green lentils (you will need 1¾ cups cooked lentils)
1 cup walnut pieces
1½ teaspoons dried oregano
1½ teaspoons ground cumin
1½ teaspoons chili powder
½ teaspoon fine-grain sea salt or to taste
1½ tablespoons extra-virgin olive oil
2 tablespoons water

TOPPINGS/WRAPS
1 tablespoon of coconut oil
1 to 2 bell peppers, thinly sliced (I used one, but next time I will use 2 for ample leftovers)
½ to 1 large onion, thinly sliced (I used ½ onion, but next time I will use 1 for ample leftovers)
Cashew Sour Cream (recipe follows)
Diced tomatoes or salsa
Green onion and fresh lime juice, for garnish
Lettuce wraps (large romaine, iceberg, or butter lettuce leaves)
Other topping options include sliced avocado, hot sauce, cilantro, etc.

Rinse lentils in a fine-mesh sieve. Add to a medium pot along with a few cups of water. Bring to a boil, reduce heat to medium, and simmer for 20 to 25 minutes or until tender (cook time will vary depending on the type of lentils you use). Drain off excess water.

Preheat the oven to 300°F. Place walnuts on a rimmed baking sheet and toast for 10 to 13 minutes, watching closely, until lightly golden and fragrant. Set aside to cool for a few minutes.

Add ½ to 1 tablespoon of coconut oil to a large skillet or wok. Cook the onion and peppers over medium heat for about 15 to 20 minutes, reducing heat if necessary and stirring frequently, until translucent.

Place 1¾ cups cooked lentils (you'll have some left over) and all the toasted walnuts into a food processor and pulse until chopped (make sure to leave texture). Stir or pulse in the oregano, cumin, chili powder, and salt. Stir in the oil and the water and mix until combined.

Prepare the rest of your vegetable toppings and wash and dry the lettuce wraps.

Assemble individual wraps with taco meat and the rest of your desired toppings.

Adapted from Oh She Glows

Cashew Sour Cream

1 cup raw cashews
2 teaspoons cider vinegar
1 teaspoon lemon juice
⅛ teaspoon fine sea salt

Place cashews in a cup or small bowl and cover by about ½ inch with boiling water. Let soak for 30 minutes. Drain cashews and place in a blender with vinegar, lemon, salt, and about ¼ cup water. Blend until very smooth, adding more water as required to purée the mixture.

Blackened Cajun Salmon

Salmon is filled with good-fat omega-3 fatty acids that promote antioxidants and have antidepressant, antiaging, and anti-arthritis effects. It's also delicious and incredibly versatile. You don't have to wait for Mardi Gras to spice meals up. (This recipe can easily be doubled or tripled.)

1 tablespoon coconut oil
2 salmon fillets or steelhead trout

CAJUN SPICE RUB
½ teaspoon oregano (fresh or dried)
½ teaspoon thyme (fresh or dried)
¼ teaspoon cayenne
¼ teaspoon smoked paprika
¼ teaspoon onion salt
¼ teaspoon garlic salt
¼ teaspoon black pepper

Melt coconut oil in a skillet or on a grill pan on medium heat. Mix spices together and spread them out on a plate. Dip both sides of your salmon fillets in this spice mix. When the pan is nice and hot, add the salmon. Turn down the heat.

Cooking time will vary depending on the thickness of your salmon. For thin pieces, start with 2 minute per side; for thicker fillets, go with 3 or 4.

Serve with sweet potatoes and a green vegetable and you'll get both glowing skin and glowing compliments from your guests.

Black Cod with Miso

Serves 2

> 1 tablespoon olive oil
> 3 tablespoons tamari
> ½ cup white miso paste
> 1 tablespoon erythritol or few drops of stevia (optional)
> 1 pound (2 to 3 fillets) black cod

Mix the olive oil, white miso paste, tamari, and sweetener (if you're using it) in a container and set aside. Clean the fillets and pat them dry. Place the fish into the container, coat them with the marinade, cover, and refrigerate overnight.

Preheat the oven to 400°F. Remove the fish from the fridge and scrape off the marinade. Coat a grill or grill pan with olive oil and set to high heat. Add the fish and cook until browned on each side, about 2 minutes.

Transfer the fillets to the oven and bake for about 10 minutes, until nice and flaky.

Braised Turmeric Cinnamon Chicken

Serves 4 to 6

> 1 whole chicken, chopped into 8 pieces
> Sea salt
> Freshly ground pepper
> 1 teaspoon ground cinnamon
> 1 tablespoon turmeric
> Olive oil
> 1 medium to large yellow onion, chopped
> 4 cloves garlic, chopped
> 2 cinnamon sticks
> 2 14-ounce cans whole peeled Italian tomatoes
> ½ cup chicken broth (or bone broth if you have it!)
> Fresh mint and parsley to garnish

Wash and dry chicken. Season with salt, pepper, and a light sprinkling of ground cinnamon and turmeric on each side.

Coat large pot with olive oil and place over high heat. When oil is hot, sear chicken pieces for 1 minute on each side, until the skins are browned. Remove chicken pieces from pan and set aside.

Lower heat to medium-high and add onions. Stir for a minute until soft, then add garlic. Let cook for another minute until translucent. Add cinnamon sticks, tomatoes, and broth and season with salt and pepper. Stir and bring to simmer. Add chicken pieces back into the pot, submerging in the liquid. Simmer for about 2 hours uncovered, shaking the pan from time to time to move the chicken around, until meat is falling off the bone.

Garnish with mint and/or parsley and serve over riced cauliflower and steamed spinach.

Pro tip: It's all about the 2-hour simmer. Slow your roll with this one. Slow food!

Grass-Fed Beef and Vegetable Stew

2 pounds stew meat from grass-fed beef
1 large sweet onion
5 large carrots
5 to 7 stalks celery
1 pound yams or butternut squash
8 cloves of garlic
3 tablespoons coconut oil (expeller-pressed)
1 cup red wine (preferably organic)
1 to 2 tablespoons organic tomato paste
6 bay leaves
3 sprigs fresh thyme
1 sprig fresh rosemary or 1 teaspoon dried rosemary (more or less to suit your taste)
½ teaspoon smoked paprika
2 quarts beef stock (homemade is best)
Sea salt and pepper to taste

Cut your stew meat into bite-size pieces. Set aside.

Chop your onions, celery, carrots, and yams or squash into bite-size pieces. Set aside. Mince garlic.

In a heavy stockpot, heat the coconut oil over medium-high heat. Add your garlic and meat and cook until the meat is browned, but be careful not to burn the garlic. Add the vegetables and stir until they are mixed in well with the meat (you may need to add a tad more oil). Add the red wine and cook for 5 to 8 minutes to allow the alcohol to cook off. Add the tomato paste and spices. Stir to combine. Add the beef stock.

Cover and bring to a simmer, and then lower the heat. Let simmer for 1 hour and then taste for salt and seasoning. Adjust the seasoning to suit your tastes. If you want a thicker stew, you can add in arrowroot at this time. The stew can be eaten at this point (provided the veggies are done), but it tastes best if it can simmer on very low heat for 3 to 4 hours before serving.

DESSERT

No-Bake Coconut Love Bites

3 cups unsweetened, shredded coconut
6 tablespoons coconut oil
½ cup xylitol or erythritol
2 teaspoons vanilla extract (I recommend fresh vanilla bean or nonalcoholic vanilla extract, since you will not be baking)
½ teaspoon sea salt
Optional toppings: shredded coconut, cocoa or carob powder, finely chopped nuts, 80 percent dark chocolate, melted, for drizzling

Put all ingredients (except optional toppings) in a food processor or blender. Combine until the mixture is blended and sticks together. (Note: If you are using a high-powered blender like a Vitamix, do not turn your machine on high.) Remove the mixture from the blender/food processor and form into desired shape. I usually make balls with a melon scooper.

Decorate with shredded coconut, cocoa or carob powder, chopped nuts, or melted chocolate as desired. I use a plastic bag with a tiny hole cut in the corner to pipe them on. You can also leave them plain.

Leave to firm up at room temperature on a plate or other hard surface.

Dark Chocolate Coconut Pudding

2 cups coconut milk

3–4 ounces dark chocolate (80 percent cacao or higher), chopped into small pieces

1 tablespoon high-quality gelatin (a form of collagen that is soluble only in hot water)

½ teaspoon vanilla extract

Pinch of sea salt

Heat the coconut milk on medium-low in a heavy-bottomed pot. Add the dark chocolate and whisk constantly until it melts. In a separate saucepan, dissolve the gelatin but do not boil.

Once the chocolate has melted, whisk in the gelatin by slowly pouring as you whisk. (If you just dump the whole tablespoon in, it'll get clumpy.) Turn off the heat and whisk in the vanilla extract.

Pour into your desired bowls or cups and chill for at least 2 hours or until set. Add salt, to taste.

You have about twenty-four thousand protein-coding genes in your body, and this reference guide contains only a small number, the ones that are mentioned in this book. Use the reference guide to look up genes and remind yourself of the names, abbreviations, and functions of each gene.

Gene, Official Name, and Function

Alzheimer's and Bad Heart Gene

Official Name: Apolipoprotein E (APOE)

Function: APOE gene instructs cells to make a protein called apolipoprotein A, which combines with fat in the body to make a package that carries cholesterol back to the liver for disposal through feces. APOE is polymorphic with three main alleles: APOEZ, APOE3, and APOE4.

Antioxidants

Official Names:

Glutathione S-transferase mu 1 (GSTM1), a gene that codes for glutathione.

Glutathione peroxidase 1 (GPX1), which detoxifies hydrogen peroxide, a reactive oxygen species.

Superoxide dismutase 2 (SOD2, or sometimes MnSOD, for manganese-dependent superoxide dismutase), helps heal mitochondria from oxidative stress.

Catalase (CAT), a gene that protects you from oxidative damage.

NAD(P)H dehydrogenase, quinone 1 (NQO1) involves coenzyme Q10

Function: Code for genes that fight oxidative damage and thereby slow down aging and prevent cancer, Alzheimer's disease, and liver damage.

Bliss

Official Name: Fatty acid amide hydrolase (FAAH)

Function: Codes for the enzyme that acts on anandamide, our natural cannabinoid molecule of bliss.

Blood Sugar and Diabetes

Official Names:

Glucose-6-phosphatase, catalytic, 2 (G6PC2)

Transcription factor 7-like 2 (TCF7L2)

Solute carrier family 30 (zinc transporter), member 8 (SLC30A8)

Hepatic lipase (LIPC)

Many others

Function: Many genes code for blood sugar, and having one or more gene variations does not mean you will have elevated blood sugar. However, you may be at greater risk of high blood sugar (fasting and after eating), caused by insulin resistance.

Brain

Official Names:

Brain-derived neurotrophic factor (BDNF)

Fatty acid amide hydrolase (FAAH)

Klotho

Amyloid precursor protein (APP)

many others

Function: Various

Breast Cancer

Official Names:

Breast and ovarian cancer susceptibility protein 1 and 2 (BRCA1 and BRCA2)

Tumor suppressor protein p53 (TP53)

Phosphatase and tensin homolog (PTEN)

Checkpoint kinase 2 (CHEK2)

ATM serine/threonine kinase (ATM)

Partner and localizer of BRCA2 (PALB2)

Many others

Function: BRCA genes belong to a class of tumor-suppressor genes that repair cell damage and keep breast cells growing normally. The TP53 gene codes for the tumor suppressor protein p53, which also regulates cell division by keeping cells from growing too fast or in an uncontrolled manner. There are at least another 100 breast cancer genes besides those mentioned here.

Caffeine Metabolism

Official Name: Cytochrome p450, family 1, subfamily a, polypeptide 2 (CYP1A2)

Function: Codes for an enzyme that breaks down caffeine and other chemicals. More than half the population are "slow metabolizers" and cannot tolerate more than 200 mg of caffeine without side effects.

Clock

Official Name: Circadian Locomotor Output Cycles Kaput (Clock)

Function: Control circadian rhythm, or the 24-hour biological sleep-wake cycle. If you have the bad variation of the gene, you will have higher blood ghrelin levels (the hormone that makes you hungry) and resistance to weight loss. Other hormones released on a circadian clock will be affected.

Corporate Warrior

Official Name: Catechol-O-methyltransferase (COMT)

Function: Keeps you focused under stress by inactivating certain brain neurotransmitters, including dopamine, epinephrine, and norepinephrine. So the normal variant makes you a corporate warrior. The polymorphism makes you a worrier, although there are potential benefits to both strategies. COMT metabolizes certain estrogens, meaning you may hang on to estrogen and increase your risk of breast cancer. Also involved in pain perception.

Deep Sleep

Official Name: Adenosine deaminase (ADA)

Function: Regulates an enzyme (also called adenosine deaminase) that converts a compound called adenosine into another compound called inosine. Adenosine is important for controlling sleep. The typical alleles are associated with less deep sleep, and the variants are linked to more deep sleep.

Detoxification

Official Names:

Methylenetetrahydrofolate reductase (MTHFR) to make usable B9 and detoxify alcohol.

Epoxide hydrolase (EPHX)

Glutathione S-transferase mu 1 (GSTM1)

Others such as CRP, CYP1A1, CYP1B1, CYP2A6, Mold (HLA DR), MMP1

Function: Helps you detoxify chemicals, toxins, and endocrine disrupters.

Eating Behaviors

Official Names:

Adrenergic beta-2 surface receptor gene (ADRB2)

Ankyrin repeat and kinase domain containing 1 (ANKK1/DRD2, or Food Desire; can influence dopamine activity and is closely linked to dopamine receptor D2 gene expression)

Fat mass and obesity associated (FTO, or Fatso)

Melanocortin 4 receptor (MC4R, or Oversnacker)

Solute carrier family 2, facilitated glucose transporter member 1 (SLCA2, or Sweet tooth)

Function: Various

Exercise

Official Names:

Peroxisome proliferator-activated receptor delta (PPARδ)

Lipoprotein lipase (LPL)

Hepatic lipase (LIPC)

Others such as MMP3, PPARGC-1-alpha, PDK4

Function: Various

Fatso

Official Name: Fat mass and obesity associated (FTO)

Function: This gene is strongly associated with your body mass index (BMI) and, consequently, your risk for obesity and diabetes. When you have the variant, it gives you sloppy control of leptin, a hormone in charge of satiety. In other words, you're hungry all the time.

High Blood Pressure

Official Name: Endothelin-1 (EDN1)

Function: Codes for endothelin-1, a potent constrictor of blood vessels. My variant of gene EDN1 puts me at a greater risk of high blood pressure if I'm inactive.

Longevity

Official Names:

Mechanistic target of rapamycin or mammalian target of rapamycin (mTOR)

Sirtuin (SIRT1)

Forkhead/winged helix box gene, group O3 (FOXO3)

Function: These genes govern longevity and autophagy (the normal physiological process of cell turnover and destruction)

Methylation
Official Names:

Methylenetetrahydrofolate reductase (MTHFR, including C677T and A1298C)

Cystathionine beta synthase (CBS)

Catechol-O-methyltransferase (COMT)

Others such as MTR, MTRR, VDR

Function: These genes—over a dozen—contribute to the methylation cycle in the body. Reminder: methylation is when a methyl group binds to a gene and ultimately may change gene expression.

Obesity, Weight Gain, Weight Loss and Regain
Official Names:

Adiponectin (various)

ADRB2

FTO

Others such as ADIPOQ (Weight loss/regain), APOA2, APOA5, GNPDA2, MC4R, PCSK1

Function: These genes contribute to obesity and fat gain when they interact with too much food, poor food choices, and too little exercise.

Seafood
Official Name: Peroxisome proliferator-activated receptor gamma (PPARγ)

Function: PPARG regulates fat cells and is involved in the development of obesity, diabetes, cancer, and heart disease. When it's turned off because you inherited the variant, you want to turn it back on so that you can lose weight by properly processing fats. Otherwise, you're at risk for a higher BMI.

Short Sleep
Official Name: hDEC2-P385R (or DEC2 for short)

Function: This gene polymorphism is linked with short sleep and resistance to sleep deprivation on fewer than 6 hours of sleep per night. Only 3 percent of the population has it.

Skin and Wrinkles
Official Names:

Pyrroline-5-carboxylate reductase 1 (PYCR1)

Matrix metalloproteinase; regulates calcium signals and collagen breakdown (MMP1)

1,500 other genes

Function: These genes determine how long you stay free of wrinkles. When you get the normal variant, your collagen stays young and healthy.

Sprinter

Official Name: Actinin alpha 3 (ACTN3)

Function: Codes for making a protein, actin, found in fast-twitch muscle fibers that allows for more explosive movement.

Stress

Official Names:

FK506 binding protein 5 (FKBP5)

Cytochrome p450, family 1, subfamily a, polypeptide 2 (CYP1A2)

Mineralocorticoid receptor (MR)

Tyrosine hydroxylase (TH)

Kidney and brain expressed protein (KIBRA) or WW domain-containing protein 1 (WWC1)

Function: Several genes regulate your stress-response system, including the amygdala, hypothalamus, pituitary, and hippocampus, the part of your brain that regulates emotion, memory, and the autonomic nervous system. Other genes regulate the way the brain talks to the adrenal glands, where cortisol is produced.

Weight Gain (see Obesity)

Vitamin D

Official Names:

Vitamin D receptor (VDR)

Others, such as vitamin D 25-hydroxylase, Fok1, Taql, CYP2R1

Function: When it's turned on, VDR codes for the structure and function of the nuclear hormone receptor for vitamin D_3, which allows your cells to absorb vitamin D. When it's turned off, you are more likely to suffer from osteoporosis.

If you want to know if you have the normal copy or a polymorphism of a gene, consider genetic testing but with the caveats mentioned in this book concerning accuracy and privacy. The following genetic variations are reported in 23andMe.com. If you decide that you want to perform genetic testing or you have already gotten your genotype, this section will help you determine what might be helpful next steps.

Companies such as 23andMe.com are limited in the information that they can provide due to sanctions from the FDA, but the reports are inexpensive (about $199 at the time of writing) and provide the raw data. To determine additional results that are easier to interpret, upload your raw data from 23andMe.com to Promethease.com or MTHFRSupport. com. These secondary services are inexpensive ways to learn more about traits like disease susceptibility.

Note that sometimes the base-pair letters are transposed depending on the orientation of the gene in the lab you're using—sometimes genes are read in the forward direction, and sometimes in the reverse direction, depending on the convention of the laboratory.[1] For instance, G=C; A=T. GG is equivalent to CC. If your dad is homozygous for the C677T mutation of MTHFR, 23andMe would report it as rs1801133 AA. See more details on reading reports from the 23andMe website.

Gene, Official Name/SNPs, and What to Do

Alzheimer's and Bad Heart Gene

Official Names/SNPs:

Apolipoprotein E (APOE) is a more complex gene because of multiple variations of two SNPs, rs429358 and rs7412. There are four alleles but one is rare (E1). The most common gene is APOE3/3, in which you inherit the APOE3 allele from each parent (called a genoset or combination of SNPs, gs246). Here are the six common patterns of inheritance.

Gene	rs429358	rs7412	Genoset	Comment
APOE2/2 Alzheimer's	(T;T)	(T;T)	gs268	Homozygous; decreased risk of
APOE2/3	(T;T)	(C;T)	gs269	
APOE2/4	(C,T)	(C;T)	gs270	
APOE3/3	(T;T)	(C;C)	gs246	Normal, most common
APOE3/4	(C;T)	(C;C)	gs141	
APOE4/4 Alzheimer's	(C;C)	(C;C)	gs216	Homozygous; increased risk of

25% of people have APOE4, which doubles or triples their risk of Alzheimer's

What to Do: If you have one or two copies of the APOE4 allele (that is, you're heterozygous or homozygous), follow the instructions in chapter 11. Most important:

- Optimize diet: low-carb, low- or no-grain.
- Fast 12 to 18 hours overnight.
- Sleep 7 to 8.5 hours/day.
- Exercise 30 to 60 minutes a day, 4 to 6 times a week (150 minutes minimum).
- Reduce inflammation (CRP < 1, homocysteine < 7).
- Reduce stress; stimulate brain.

Breast Cancer

Official Names/SNPs:

BRCA1 (at least 122 SNPs)

BRCA2 (at least 129 SNPs)

What to Do: If you have a variant that increases your risk of breast cancer, consider these actions:

- Weight loss if BMI ≥ 25.
- Drink < 3 servings alcohol/week or abstain.
- Get regular breast-cancer screening.
- Possible medication to reduce risk (tamoxifen, raloxifene, aromatase inhibitors).
- Possible prophylactic surgery, if appropriate (removal of breasts and/or ovaries).

Clock

Official Names/SNPs:

Circadian Locomotor Output Cycles Kaput (Clock) / rs1801260

Normal (C;C)

Heterozygous (C;T)

Homozygous (T;T)

What to Do:

- Variants have higher blood ghrelin levels, the hunger hormones, and weight loss resistance.
- Get 8 hours of sleep every night in order to lose weight.
- Keep circadian clock as regular as possible with consistent sleep/wake cycle every day.

Fatso

Official Names/SNPs:

Fat mass and obesity associated (FTO) / rs9939609

Normal (T;T)

Heterozygous (A;T) has 1.3 times the risk for type 2 diabetes and an increased obesity risk

Homozygous (A;A) has 3 times obesity risk and 1.6 times risk for type 2 diabetes

What to Do:

- You are at greater risk of obesity if eating poorly.
- Track fasting glucose and hemoglobin A1c, reduce size of carb portion at meals.
- Exercise and low-carb diet help.

Longevity

Official Names/SNPs: Mechanistic target of rapamycin or mammalian target of rapamycin (mTOR) / multiple SNPs

What to Do:

- Turn off mTOR with intermittent fasting, nutritional ketosis, healthy fats. Additional supplements that help:
  - Di-indole methane (DIM)
  - N-acetylcysteine
  - Resveratrol
  - Lipoic acid

Official Names/SNPs: Sirtuin (SIRT1) / multiple SNPs

What to Do:

- Similarly, turn on SIRT1 with intermittent fasting, nutritional ketosis, healthy fats.
- Specifically, increase DHA by eating cold-water fish or take a supplement.
- Tighten blood-sugar control (keep fasting blood sugar 70 to 85 mg/dL, and 2-hour postprandial < 120 mg/dL).
- Bathe in a dry or infrared sauna.
- Exercise regularly, especially burst exercise or adaptive exercise (yoga, Pilates, tai chi).
- Reduce oxidative stress.

Official Names/SNPs:

Forkhead/winged helix box gene, group O3 (FOXO3) / rs2802292 (plus multiple other SNPs)

Normal (T;T)

Heterozygous (G;T) is associated with between 1.5 to 2 times increased likelihood of living to be a centenarian.

Homozygousr (G;G) is associated with between 1.5 to 2.7 times increased likelihood of living to be a centenarian.

What to Do: Sit in a hot, dry sauna for 20 minutes at least four times per week to turn on this gene.

Methylation

Official Names/SNPs:

Methylenetetrahydrofolate reductase (MTHFR) / rs1801133 (there are several)

Normal (G;G)

Heterozygous (A;G) has 35 to 40% reduced MTHFR enzyme activity.

Homozygous (A;A) has 80 to 90% reduced MTHFR enzyme activity.

What to Do: If you have a variant of one or more of these SNPs, you may have an elevated homocysteine level in your blood, and low vitamin B12 and folate. You have a higher probability of poorly processing folic acid. See your functional medicine professional for doses based on your methylation activity and context. Consider supplementing with 5-methylfolate (or L5MTHF), methylcobalamin (vitamin B_{12}), and riboflavin to work around the gene; track homocysteine in the blood.

Vitamin D Receptor

Official Names/SNPs:

VDR / rs1544410

Normal (T;T)

Heterozygous (G;T)

Homozygous (G;G)

What to Do:

* Keep vitamin D 60 to 90 ng/mL for optimal healthspan.

Adiponectin. Also known as apM1, AdipoQ, Acrp30, and GBP-28. Adiponectin is encoded by the ADIPOQ gene and is secreted by fat cells. It regulates glucose levels and fat burning.

Adrenal glands. Endocrine glands that produce hormones, such as sex hormones and cortisol, which help you respond to stress and have many other functions. You have an adrenal, or suprarenal, gland at the top of each kidney.

Adrenocorticotropic hormone (ACTH). Also known as corticotropin. A hormone released from the anterior pituitary gland in the brain, ACTH is an important component of the hypothalamic-pituitary-adrenal axis because it increases production of cortisol in the adrenal gland. Produced in response to stress, ACTH levels in the blood are measured to help detect, diagnose, and monitor conditions associated with excessive or deficient cortisol in the body.

Allele. An allele is a variant form of a gene. For each genetic locus in your chromosome, you have two alleles. You inherit one allele (copy of a gene) from your mother and one copy from your father. If the alleles you inherit are the same, you are homozygous for that gene. If the alleles are different, you are heterozygous for that gene.

Amyloid beta. Sticky peptides, or groups of amino acids, that can aggregate together and form amyloid plaques. The peptides come from a larger precursor protein (amyloid precursor protein, or APP), which is cleaved to yield amyloid beta. It impairs the structure and function of your tissues, collects in the brain, and is toxic to nerve cells, increasing the risk for Alzheimer's disease.

Beta amyloid. See amyloid beta.

Brain-derived neurotrophic factor (BDNF). Part of the family of proteins known as neurotrophic factors that contribute to the growth and survival of nerve cells. BDNF is found in the brain and spinal cord and is active at the connections between nerve cells, known as synapses. BDNF promotes synaptic plasticity, facilitates neural repair, and enhances learning and memory.

Collagen. An easily digested form of protein that improves skin, hair, and nails. As you age, you break down more internal collagen than you make, leading to saggy skin, cracking fingernails, dull hair, and wrinkles.

Corticotropin-releasing hormone (CRH). A hormone involved in the stress response system. It is secreted by the hypothalamus and stimulates the pituitary gland to make adrenocorticotropic hormone (ACTH). Excess stress and over exercising raise CRH levels, which can increase the permeability of the intestinal walls as well as the permeability of the lungs, skin, and blood-brain barrier. CRH can also be released outside of the central nervous system, such as in the skin, where it may cause inflammation.

Deoxyribonucleic acid (DNA). A repeating pattern of four chemical bases: adenine (A), cytosine (C), guanine (G), and thymine (T). These bases are the alphabet in which your genetic code is written. The bases pair up with each other—A with T, and C with G—to form a base pair. Your DNA is like a ladder, with the base pairs forming the ladder's rungs. (The sides of the ladder are composed of sugar and phosphate.) You have three billion bases in your genome, and 99.5 percent of them are the same from human to human.

Epigenetics. Refers to the changes in gene expression caused by mechanisms other than the DNA sequence. Certain triggers may override your gene expression, silencing a bad gene or promoting a good gene.

Epinephrine. A hormone with neurotransmitters made in the inner core of the adrenals that help you focus and problem-solve. It creates amounts of glucose and fatty acids that can be used by the body as fuel in times of stress or danger when increased alertness or exertion is required.

Gene. Your genes are collections of base pairs that give the recipes to make specific proteins, such as enzymes. Each gene makes about three proteins. The sequence of the bases tells your body how to build, repair, and maintain itself. You inherit one copy of a gene from your mother and one copy from your father. The single copies are called alleles. If you inherit the normal copy of the gene from each parent, you are normal, or wild type. If you inherit one copy of the normal gene and one copy of the polymorphism, you are heterozygous. If you inherit two copies of the same polymorphism, you are homozygous. Problems arise mostly when you are heterozygous or homozygous.

Gene regulation. Refers to the mechanisms used by the cell to control which genes are expressed and to increase or decrease the production of RNA and proteins.

Genetics. Refers to the function and composition of specific genes.

Genomics. Refers to how all your genes are expressed in your body.

Glucocorticoids. Made in the outside portion (the cortex) of the adrenal gland, glucocorticoids regulate the metabolism of glucose and are chemically classed as steroids. Cortisol is the major natural glucocorticoid.

Hypothalamic-pituitary-adrenal (HPA) axis. A feedback loop by which signals from the brain trigger the release of hormones needed to respond to stress. Because of its function, the HPA axis is also sometimes called the stress circuit.

Insulin. Drives glucose into cells as fuel and deposits fat. Chronically high insulin increases estrogen, and estrone, specifically, increases the cells' resistance to insulin.

Irisin. A hormone secreted from muscles in response to exercise. It tricks white fat into behaving like brown fat, builds muscle, activates weight loss, and blocks diabetes.

Leptin. A hormone that controls hunger, metabolism, and the utilization of food as fuel or fat.

Maximal heart rate. The highest heart rate you can achieve during maximum physical exertion. To calculate, subtract your age from 220. This is the maximum number of times your heart should beat per minute while you are exercising.

Melatonin. A hormone secreted by the pineal gland in the brain that helps regulate other hormones and maintains the body's circadian rhythm. Melatonin also helps control the timing and release of female reproductive hormones.

Methylfolate. The enzyme methylenetetrahydrofolate reductase (MTHFR) takes folate (vitamin B9) and converts it into methylfolate (L5MTHF). Activated methylfolate plays a key role in the biochemical process of methylation. Methylation is the powerhouse detoxification, production, and DNA protection system that almost every cell of your body depends on.

Mitochondrial dysfunction. Occurs when the mitochondria are not able to perform their job and is one of the cellular signs of aging. Causes include nutritional deficiencies and excesses, toxin exposure, oxidative stress, and microbial infection (or dysbiosis). Tired mitochondria may make you feel more fatigued during and after exercise or cause muscle pain.

Myokines. Small proteins released when your muscles are contracting. These proteins enter the bloodstream and their levels increase before and after exercise. Higher levels of myokines in skin cells have been associated with younger-looking skin.

Myostatin. A growth factor that regulates muscle size and prevents them from growing too large. A lack of myostatin results in excessive muscle growth. It may also control loss of muscle mass in aging women.

Nerve growth factor. A neurotrophic factor (part of the same family of proteins as BDNF) and neuropeptide. It regulates the growth, maintenance, proliferation, and survival of certain neurons. Yoga is associated with higher levels of nerve growth factor.

Norepinephrine. A neurotransmitter made in the inner core of the adrenals that helps with focus and problem solving. It acts as a neuromodulator in the nervous system and as a hormone in the blood.

Oxidative stress. Refers to an imbalance between the production of reactive oxygen species (free radicals) and antioxidants. Free radicals are oxygen-containing molecules with one or more unpaired electrons that can interfere with and destabilize DNA, proteins, fats, and other cell components. Antioxidants neutralize and counteract the harmful effects of free radicals.

Oxytocin. Both a hormone and a neurotransmitter, which means it acts as a brain chemical that transmits information from nerve to nerve; called by some "the love hormone" because it increases in the blood with orgasm in both men and women. Oxytocin is also released when the cervix dilates, thereby augmenting labor, and when a woman's nipples are stimulated, which facilitates breast-feeding and promotes bonding between mother and baby.

Single-nucleotide polymorphism (SNP, pronounced "snip"). SNPs are slight variations in genes. The variation refers to the sequence in which a single nucleotide—building block of DNA—is changed.

Synaptoporosis. Refers to the problem Alzheimer's patients have with maintaining the balance between actively remembering and forgetting memory inputs. Amyloid precursor protein (APP) manages this process in the brain, and the delicate balance is completely impaired in Alzheimer's patients.

Thyroid. A gland that keeps the metabolism balanced, giving you energy, comfortable warmth, and manageable weight.

Transcription factors. Proteins involved in the process of converting, or transcribing, DNA into RNA. They bind to specific sequences of DNA, thereby controlling the rate of transcription.

Vagal tone. Refers to the responsiveness of the vagus nerve. Lower vagal tone means the vagus nerve is not fully performing its functions and can lead to a variety of problems. Meditation can enhance vagal tone.

Vagus nerve. The most important nerve and the portal to the parasympathetic nervous system. If your vagus nerve is impaired, you won't be healthy and are more likely to age faster.

Vasopressin (AVP). A hormone released by the hypothalamus in response to a stressful threat. It retains water in the body and constricts blood vessels.

Vitamin D. Synthesized from cholesterol and exposure to sunlight. It can also be ingested from food, but it is not officially an essential vitamin because it can be made by mammals exposed to the sun. It is considered a vitamin and a hormone. Present in eggs and fish, it is added to other foods such as milk; available as a dietary supplement.

Test your DNA (with all of the caveats mentioned earlier)

- 23andMe: Perhaps the best-known personal genetics company, 23andMe lets you begin your gene exploration online by ordering one of its kits for $199 (prices subject to change). Once you receive your Personal Genome Service, you register and spit into the included container. After you mail back your saliva, it takes six to eight weeks for your results to be processed. Once the test has been completed, you can take advantage of 23andMe's relative finder in addition to learning about your disease risk.
- Pathway: The PathwayFit test gives you a personalized look into your genetic code and analyzes your metabolism, eating habits, and the way your body responds to exercise. The test, along with a lifestyle questionnaire, tells you how to optimize your diet, workouts, and lifestyle for a strong metabolism.
- SmartDNA: SmartDNA offers genomic testing through registered practitioners. Their Genomic Wellness Test covers over a hundred DNA changes and provides a comprehensive analysis with action steps for a personalized optimal wellness program.
- Gene by Gene: The tests offered by Gene by Gene range in price from a $195 for a non-legal DNA profile to a $950 forensic-infidelity test. Its tests cater to specific circumstances, such as Paternity Peace of Mind, Twin-Zygosity, and Complex Family Reconstruction.
- DNA Ancestry: This $99 DNA test, a new service from Ancestry.com, focuses on your family's origins. The entire database boasts some ten billion entries and thirty-four million family trees. While other tests emphasize health, this one emphasizes region of family origin.

Other tests mentioned in the book

- Psychomotor Vigilance Task
  - www.buypvt.com/
  - https://itunes.apple.com/us/app/psychomotor-vigilance-test/id1034227676?mt=8
  - https://itunes.apple.com/us/app/mind-metrics/id460744094?mt=8
- Methylation panels
  - Doctor's Data methylation profile provides functional assessments of the phenotypic expression of methylation SNPs in plasma. www.doctorsdata.com/methylation
    -profile-plasma/
  - HDRI methylation pathway panel www.hdri-usa.com/tests/methylation/
- Microbiome: Ubiome http://ubiome.com, Doctor's Data Comprehensive Stool Analysis www.doctorsdata.com/comprehensive-stool-analysis/, or smartGUT by www.smartDNA.com
- Quicksilver Mercury Tri-Test measures methylmercury and inorganic mercury, allowing analysis of exposure sources, body burden, and the ability to excrete each form of mercury. Building an informed picture allows your practitioner to plan a successful approach to detoxification. www.quicksilverscientific.com/testing/mercury-tri-test

Recommended laboratory tests

Basic blood panels: Blood tests speak the language of conventional physicians, so I typically start with blood (and try to build a bridge). Ask your doctor to order the following:
- VAP cholesterol: includes subtypes of LDL and HDL plus lipo(a), VLDL
- Ferritin
- Thyroid panel: TSH, free T3, reverse T3
- Adrenal panel: cortisol, DHEA
- Sex hormones: estradiol, progesterone, DHEA; free, bioavailable, and total testosterone
- Liver function (ALT, AST, total bilirubin)
- Fasting blood glucose
- Hemoglobin A1c
- Homocysteine
- High-sensitivity C-reactive protein (hsCRP)
- If you're overweight, add leptin, insulin, IGF-1 (growth hormone).
- If your doctor won't order these tests, consider going to WellnessFx.com or MyMedLab.com.

Omega-6/omega-3 ratio: Offered from Metagenics or Vital Choice. You can also get this done as part of the NutrEval (see below) by Genova. If you are having new symptoms of ADD in perimenopause, get this test. Omega-3s have been proven to be effective, yet most people don't optimize their levels.

Additional hormone profile: If your doctor is the more open-minded type, check out one of the following tests.
- Dried Urine Test for Comprehensive Hormones by Precision Analytical. It will tell you about your adrenal health and inform you of your estrogen metabolism (that is, if you have a modifiable tendency toward breast cancer). https://dutchtest.com
- Genova's Complete Hormones test. www.gdx.net/product/complete-hormones -test-urine
- Genova's Menopause Plus will test your melatonin and cortisol levels in your saliva, as well as your estrogen and progesterone. What I like about this one is that they'll test your estrogen and progesterone over three days for a more representative result. www.gdx.net/product/menopause-plus-hormone-test-saliva

Genova's NutrEval: For those of you who really love to measure everything, and want to know where your nutritional gaps and excesses are, consider your prayers answered: www.gdx.net/product /10051. The cost is reasonable for people with insurance who qualify for Genova's Pay Assured program at $169. Add on vitamin D for $5.

Heavy metals:
- Doctor's Data unprovoked or provoked test for urine toxic metals. www.doctorsdata .com/urine-toxic-metals/
- Mercury. I commonly see women with fatigue, hair loss, weight gain, low sex drive, and underperforming thyroids. I encourage them to take Quicksilver's Tri-Test for mercury. www.quicksilverscientific.com/testing/mercury-tri-test

Telomeres: For those soul sisters who delight in quantifying biological age, the best marker is your telomeres, the little caps on the ends of chromosomes, similar in function to the aglets on shoelaces, that keep your chromosomes from unraveling. Get those telomeres tested at Lifeline.com (best) or Spectracell.com.

Food sensitivity testing: Cyrex provides multi-tissue antibody testing for the early detection and monitoring of the complex autoimmune conditions that may be accelerating the aging process for you. Array 2 will assess leaky gut, array 3 checks for gluten sensitivity, and array 4 will look at cross-reactive food sensitivities. www.cyrexlabs.com

Ways to calm down with your smartphone

- 10% Happier. My favorite app for wandering minds.
- Calm. This is a simple mindfulness meditation app that will bring more joy, clarity, and peace of mind into your life.
- Headspace. This app will help you learn mindfulness meditation in just ten minutes a day! Free for ten days.
- Insight Meditation Timer. Download the app and join my group Younger (search for *Younger* in the groups). Meditate with us and post us a message on your favorite visualizations and other practices.

Life coaching

Handel Group www.handelgroup.com
New Ventures West www.newventureswest.com

Drinks

ENERGIZING TEA VARIETALS

Green tea is loaded with antioxidants and nutrients. It is used to improve mental alertness and thinking and has a host of health benefits, including preventing atherosclerosis, lowering high cholesterol, and controlling blood-sugar levels. Here are some of my favorite brands:

- Tealux
- Stash organic
- Genmaicha Japanese Loose Leaf green tea (this is my favorite green tea)
- Matcha is green tea that has been finely ground into a powder.

White tea is made from buds and young leaves, and is minimally processed. It is loaded with polyphenols and is an anti-inflammatory and antioxidant.

- Tealux (this white tea has been shown to be low in lead)

Oolong tea is a traditional Chinese tea made from the leaves, buds, and stems of the *Camellia sinensis* plant. It is used to improve mental clarity.

- Numi's Iron Goddess Oolong tea is one of my favorites. It's light and delicately flavorful with a smooth body and sweet finish.

COFFEE

- Bulletproof coffee is grown at high altitudes in Central America, hand-harvested, and carefully processed to minimize performance-robbing mold toxins and to maintain flavor.
- Longevity from David Wolfe is rich in antioxidants and has a robust, delicious flavor.

COLLAGEN (FOR COLLAGEN LATTE, PAGE 85)

- Bulletproof Collagen Protein is sourced from pasture-raised cows untouched by drugs or hormones. Benefits include increased energy, speedier recovery, and a boosted immune system.
- Green Lakes Hydrolyzed Collagen is also sourced from grass-fed cattle. The hydrolyzed collagen gelatin allows for rapid absorption and is soluble in cold water.

Recovery Paste

After burst training of a minimum of 4 to 5 rounds of maximal effort, replete your glycogen stores and repair your muscles with a recovery paste. Mix 2 scoops of Dr. Sara's Reset360 All-in-One Shake in chocolate (I prefer chocolate but other flavors are vanilla, berry, and cappuccino) with coconut water to desired consistency. Add stevia to desired sweetness and consume within 45 minutes of workout.

Supplements

- Berberine dose is 400 mg once or twice per day. Efficacy improved with milk thistle. Take a holiday after two months, then restart later if fasting blood sugar is elevated. Recommended: Aging Reset Essentials at Reset360.com.
- Resveratrol dose is 200 mg once per day. Recommended: Aging Reset Essentials at Reset360.com.
- Branch-chain amino acids (BCAAs) dose is 3 to 8 grams during or immediately after workout. Recommended: Designs for Health, Pure Encapsulations, and Thorne.

How to measure and reset your blood sugar

Frequency: Daily; however, measure every week if your fasting and postprandial (after-a-meal) blood-sugar levels are in the optimal ranges.

Supplies needed: A blood glucose meter (you can purchase one at your local drugstore without a prescription), blood glucose test strips, a lancing device, lancets, and a control solution (optional).

Instructions: There are two important times to check your blood sugar. The first is in the morning, after you've not eaten for eight to twelve hours, and the second is two hours after you've eaten. Start by measuring your fasting blood sugar before eating breakfast. In addition, it's helpful to measure your blood sugar two hours after a meal, particularly dinner.

Heart rate variability

If you ask your doctor what your pulse is, he or she will usually give you a single number, typically between 60 and 90 beats per minute. But the heart does not beat like an unchanging metronome; the intervals between one heartbeat and the next are variable. HRV is the beat-to-beat variation in consecutive heartbeats as shown in electrocardiogram (EKG) recordings.

So if the number your doctor gives you for your heart rate is 62, your heart is really beating between some range, such as 56 and 67. A healthy heart rate is always variable because a body is always in a state of physiological and emotional change. When you inhale, your heart rate speeds up, and when you exhale, it slows down.

HOW TO MEASURE HEART RATE VARIABILITY (HRV)

There are many options available to monitor HRV at home, on the go, and while working out. Most require an app on a device such as an iPhone and a heart rate monitor. There are wrist-based monitors, as well as Bluetooth-enabled chest-strap monitors. I recommend a chest-strap version, as it is more likely to give a clinical-grade measurement. Two apps that I recommend are SweetBeat HRV and HeartMath's Inner Balance Transformation System.

- The SweetBeat HRV app focuses on stress reduction, training, and heart rate recovery. It will alert you when your HRV is not at a healthy level, prompting you to engage in a stress-reduction effort. The training component assesses whether you can train full out, or whether you should have a low-intensity day or even a day off.
- The HeartMath Inner Balance Transformation System is an application that comes with a sensor and an earpiece that takes a pulse reading from your earlobe. The system encourages you to moderate your breathing and focus on positive emotions in order to reduce negative stress, improve relaxation, and build resilience. By synchronizing your breathing using a breathing pacer on the app, you can make your HRV healthier and achieve a state of coherence.

Favorite forms of movement

- Chi Walking and Chi Running www.chirunning.com
- Forrest Yoga was developed by Ana Forrest and is an intensely physical and internally focused practice. www.forrestyoga.com
- Barre classes are a combination of postures inspired by ballet, yoga, and Pilates. The barre is used as a prop to balance while doing exercises.
  - Dailey Method: www.thedaileymethod.com
  - Barre3: http://barre3.com
  - Bar Method: http://barmethod.com
  - Pure Barre: www.purebarre.com

RELEASE TECHNIQUES

- Active-release therapy (ART) is a soft-tissue, movement-based technique developed by P. Michael Leahy. The aim is to address problems with muscles, tendons, ligaments, fascia, and nerves. www.activerelease.com/
- The Feldenkrais Method uses gentle movement and directed attention to improve movement and enhance human function. Benefits include increased ease and range of motion, and improved flexibility and coordination. www.feldenkrais.com/
- Tension and trauma-releasing exercises (TRE). This is a technique that uses a set of six exercises to release deep tension in the body by evoking self-controlled tremors, a muscle-shaking process sometimes called "neurogenic muscle tremors." http://trauma prevention.com
- Anat Baniel Method has helped me with chronic tension in my neck, shoulders, and chest. Anat is a clinical psychologist and dancer whose focus is on how to reorganize the brain through movement. Her subtle work has helped me learn how to release old grooves of long-standing tension. www.anatbanielmethod.com
- Yamuna body rolling is a fitness method developed by Yamuna Zake. It combines healing, wellness, and injury prevention in a single, simple workout. Balls of various sizes and firmness, along with body weight and small movements, are used to release tension from your neck, back, leg muscles, and more. www.yamunausa.com/
- Jill Miller of Yoga Tune Up believes that restrictions in the diaphragm can make it hard to calm down the nervous system. She advises using small, grippy balls or a two tennis balls, and placing them under the midback.
- Other: Sue Hitzmann's Melt Method. www.meltmethod.com

Exposures

MOLD

To learn more about mold, I recommend Dr. Ritchie Shoemaker's excellent website Survivingmold.com.

SAFE SKIN PRODUCTS

- Annmarie Gianni. Annmarie Skin Care sells effective, organic, and cruelty-free beauty products made from natural oils and herbs. www.saragottfriedmd.com/skincarelove/
- Tarte Cosmetics are clean and serious powerhouses in the makeup department.
- Josie Moran cosmetics are made from natural, organic, nontoxic, and environmentally friendly ingredients. The company is committed to selling makeup that looks as good on you as it is for you. www.josiemarancosmetics.com/
- OSEA is a natural skin-care brand that provides high-quality, ecologically responsible, and natural skin care. OSEA stands for "ocean, sun, earth, and atmosphere," and the company is committed to the four elements of our planet and working with nature for the most pure and effective products. http://oseamalibu.com/

- Hairprint is committed to healthier hair-coloring alternatives based on green chemistry. They offer a nontoxic method to restore gray hair to its natural color. It works for brunettes and darker-colored hair, not blondes, although I've had good success with my light brown hair. www.myhairprint.com

SAUNAS

Saunas increase circulation, lower blood pressure, and are associated with longevity. It's like a mini workout, and the heat allows you to sweat out toxins through your skin. In my home, I have a two-person sauna by Sunlighten. My husband and I love it. Date night in the sauna! www.sunlighten.com/

TOXIN-FREE TAMPONS AND ALTERNATIVES

- Seventh Generation tampons
- Veeda tampons
- Natracare tampons
- Diva Cup menstrual cups
- Lena menstrual cups

ALTERNATIVES TO PAINT WITH VOLATILE ORGANIC COMPOUNDS (VOCS)

- low- or no-VOC paints
  - Mythic Paint
  - Colorhouse Paint
  - AFM Safecoat Paint
  - Milk Paint
- The U.S. Green Building Council is transforming the way buildings are designed, built, and operated. For more information, visit www.usgbc.org/.

Brain games

NeuroRacer is a therapeutic video game created by Professor Adam Gazzaley of the University of California at San Francisco. Gazzaley designed NeuroRacer to combat age-related mental decline by using neurofeedback and TES (transcranial electrical stimulation) to boost brainpower. The game specifically aids working memory and attention span, and these improved skills were transferable to the real world. Learn more at Professor Gazzaley's lab's website: http://gazzaleylab.ucsf.edu/neuroscience-projects/neuroracer/.

Trackers

For tracking your fitness and sleep, I recommend Jawbone Up and Misfit Ray.

Vision testing and improvement

- www.essilov.com, www.visiongym.com
- Apps: Attentive Eye Test, Vision Test, and Eye Chart

ACKNOWLEDGMENTS

Thank you to all of the beautifully aging people that I've been stalking in my hometown at Whole Foods, farmer's markets, in my barre and yoga classes, and online.

As I organized my ideas and threw myself into creative labor, many friends and colleagues helped me clarify and organize my thinking or even offered themselves as readers or cases: Dr. Betty Suh Burgmann, Dr. Alan Christianson, Ana Forrest, Betty Fussell, Kevin Gianni, Allison, Maureen, Renske, Sandy, Chris Kresser, Sylvia, Nick Polizzi, Meryl Rosofsky, and Robyn Scherr.

My extraordinary agent, Celeste Fine, astounds me regularly with her cogent and smart ideas for this book and many others that we will create in the future. I am blessed to work with her.

Once I started writing, I couldn't have done without the organization and grace brought by my luminous editorial and coaching team: Andrea Vinley Jewell, Tracy Roe, and Autumn Millhouse. Kevin Plottner kept the lid on my stress-response system with his beautiful designs for the infographics. Special thanks to my excellent team: Rachel, Molly, Laura, Yoni, and Zehava. Warmest gratitude to my comrade-in-arms Christina Wilson for coaching the online tribe! Thanks also to the extraordinary ambassadors at Gottfried Institute who help spread the message and answer questions in our online courses.

Gideon Weil is the best health editor in the industry, perhaps also the funniest, and I'm so blessed to work with him. Warmest thanks to the outstanding production, publicity, and marketing team, including Melinda Mullin, Laina Adler, Amy VanLangen, Noël Chrisman, and Terri Leonard. I'm grateful to Mark Tauber for his leadership at HarperOne and for making it all happen.

Thank you to my loving and supportive daughters. I hope this book helps you leverage epigenetics so you can work around the mix of genes I passed on to you. I'm grateful to my parents, Albert and Mary Lil Szal, for their patient responses to my perpetual questions while writing this book and for the past fifty years. The book would not have been as fun to write without the support and love of my awesome sisters, Anna and Justina.

Most of all, thanks to my two secret weapons: Johanna Ilfeld, Ph.D., my close friend and fitness buddy who read draft after draft, always with affection, humor, and smart insights; and David Gottfried, my husband, sounding board, gifted editor, life companion, soulmate, and beloved.

Introduction: Women, Aging, and Genetics

1. Rappaport, S. M. "Implications of the exposome for exposure science." *Journal of Exposure Science and Environmental Epidemiology* 21, no. 1 (2011): 5–9; Rappaport, S. M., et al. "Using the blood exposome to discover causes of disease." *Agilent Technologies,* September 15, 2015, accessed February 9, 2015, www.agilent.com/cs/library/technicaloverviews/Public/5991–3418EN.pdf; Harmon, K. "Sequencing the exposome: Researchers take a cue from genomics to decipher environmental exposure's links to disease." *Scientific American,* October 21, 2010, accessed February 2, 2016, www.scientificamerican.com/article/environmental-exposure.

2. "Vital Statistics Rapid Release." *Centers for Disease Control and Prevention,* accessed April 11, 2016, www.cdc.gov/nchs/products/vsrr/mortality-dashboard.htm; Ludwig, D. S. "Lifespan weighed down by diet." *JAMA* (2016).

3. Vincent, G. K., et al. *The Next Four Decades: The Older Population in the United States: 2010 to 2050* (U.S. Department of Commerce, Economics and Statistics Administration, U.S. Census Bureau, 2010).

4. Hebert, L. E., et al. "Annual incidence of Alzheimer disease in the United States projected to the years 2000 through 2050." *Alzheimer Disease and Associated Disorders* 15, no. 4 (2001): 169–73; Alzheimer's Association, "2015 Alzheimer's disease facts and figures." *Alzheimer's and Dementia: Journal of the Alzheimer's Association* 11, no. 3 (2015): 332.

5. "U.S. breast cancer cases expected to increase by as much as 50 percent by 2030." *American Association for Cancer Research,* accessed April 30, 2015, www.aacr.org/Newsroom/Pages/News-Release-Detail.aspx?ItemID=708#.VUJpv61VhBc; Brown, E. "Breast cancer cases in U.S. projected to rise as much as 50% by 2030." *Los Angeles Times,* April 20, 2015, accessed April 30, 2015, www.latimes.com/science/sciencenow/la-sci-sn-breast-cancer-cases-2030-20150420-story.html.

6. van Drielen, K., et al. "Disentangling the effects of circulating IGF-1, glucose, and cortisol on features of perceived age." *Age* 37, no. 3 (2015): 1–10.

7. Krøll, J. "Correlations of plasma cortisol levels, chaperone expression and mammalian longevity: a review of published data." *Biogerontology* 11, no. 4 (2010): 495–99; van Drielen, K., et al. "Disentangling the effects of circulating IGF-1, glucose, and cortisol"; Christensen, K, et al. "Perceived age as clinically useful biomarker of ageing: cohort study." *British Medical Journal* 339 (2009): b5262; Noordam, R., et al. "Serum insulin-like growth factor 1 and facial ageing: high levels associate with reduced skin wrinkling in a cross-sectional study." *British Journal of Dermatology* 168, no. 3 (2013): 533–38; Noordam, R., et al. "Cortisol serum levels in familial longevity and perceived age: the Leiden longevity study." *Psychoneuroendocrinology* 37, no. 10 (2012): 1669–75.

Chapter 1: Unlock Your Genes

1. "Cleveland Clinic's Center for Functional Medicine: A Test Kitchen for Healthcare's Future," Holistic Primary Care, https://holisticprimarycare.net/topics/topics-a-g/functional-medicine/1680-cleveland-clinic-s-center-for-functional-medicine-a-test-kitchen-for-healthcare-s-future.html; http://my.clevelandclinic.org/services/center-for-functional-medicine.

2. Gifford, B. *Spring Chicken: Stay Young Forever (or Die Trying)* (New York: Grand Central, 2015), 37. Data are from 2012.

3. "Top five cosmetic surgical procedures of 2013." *PlasticSurgery.org,* accessed April 10, 2015,

Notes **317**

www.plasticsurgery.org/news/plastic-surgery-statistics/2013/top-five-cosmetic-surgery
-procedures.html.

4. Furnham, A., et al. "Factors that motivate people to undergo cosmetic surgery." *Canadian Journal of Plastic Surgery* 20, no. 4 (2012): e47.
5. "Get your bolder on. We can all use a little motivation." GrowingBolder.com, accessed June 25, 2015, www.growingbolder.com/quotes/#.
6. Goodstein, G. "Ida Keeling still setting records, examples at 100." *Bronx Times*, May 8, 2015, accessed June 25, 3015, www.bxtimes.com/stories/2015/19/19-ida-2015-05-08-bx.html.
7. Arem, H., et al. "Leisure time physical activity and mortality: A detailed pooled analysis of the dose-response relationship." *JAMA Internal Medicine* 175, no. 6 (2015): 959-67.

Chapter 2: The Gene/Lifestyle Conversation

1. Memisoglu, A., et al. "Interaction between a peroxisome proliferator-activated receptor γ gene polymorphism and dietary fat intake in relation to body mass." *Human Molecular Genetics* 12, no. 22 (2003): 2923-29.
2. Walsh, T., et al. "Ten genes for inherited breast cancer." *Cancer Cell* 11, no. 2 (2007): 103-5; Walsh, T., et al. "Spectrum of mutations in BRCA1, BRCA2, CHEK2, and TP53 in families at high risk of breast cancer." *JAMA* 295, no. 12 (2006): 1379-88; Aloraifi, F., et al. "Gene analysis techniques and susceptibility gene discovery in non-BRCA1/BRCA2 familial breast cancer." *Surgical Oncology* 24, no. 2 (2015): 100-109; Lee, D. S. C., et al. "Comparable frequency of BRCA1, BRCA2 and TP53 germline mutations in a multi-ethnic Asian cohort suggests TP53 screening should be offered together with BRCA1/2 screening to early-onset breast cancer patients." *Breast Cancer Research* 14, no. 2 (2012): R66. For lay audiences, these citations may be helpful: "Genetics," *BreastCancer.org*, accessed February 13, 2016, www.breastcancer.org/risk/factors/genetics; "Inherited gene mutations," *Komen.org*, accessed February 13, 2016, http://ww5.komen.org/BreastCancer/InheritedGenetic Mutations.html; "Breast cancer genes," *Cancer Research UK*, accessed February 13, 2016, www.cancerresearchuk.org/about-cancer/type/breast-cancer/about/risks/breast -cancer-genes.
3. Winkler, T. W., et al. "The influence of age and sex on genetic associations with adult body size and shape: A large-scale genome-wide interaction study." *PLoS Genetics* 11, no. 10 (2015): e1005378.
4. "Orientation," *SNPedia*, August 15, 2015, accessed October 20, 2015, http://snpedia.com /index.php/Orientation.
5. Miller, J. W., et al. "Vitamin D status and rates of cognitive decline in a multiethnic cohort of older adults." *JAMA Neurology* (2015); Wilson, V. K., et al. "Relationship between 25-hydroxyvitamin D and cognitive function in older adults: The health, aging and body composition study." *Journal of the American Geriatrics Society* 62, no. 4 (2014): 636-41; Chei, C. L., et al. "Vitamin D levels and cognition in elderly adults in China." *Journal of the American Geriatrics Society* 62, no. 11 (2014): 2125-29; Littlejohns, T. J., et al. "Vitamin D and the risk of dementia and Alzheimer disease." *Neurology* 83, no. 10 (2014): 920-28; Annweiler, C., et al. "Vitamin D-mentia: randomized clinical trials should be the next step." *Neuroepidemiology* 37, nos. 3-4 (2011): 249-58.

Chapter 3: Epigenetics: Turning Genes On and Off

1. Siddhartha Mukherjee, *The Gene: An Intimate History* (New York: Scribner: 2016), 400.
2. Audergon, P., et al. "Restricted epigenetic inheritance of H3K9 methylation." *Science* 348, no. 6230 (2015): 132-35.
3. Shapira, I., et al. "Evolving concepts: How diet and the intestinal microbiome act as modulators of breast malignancy." *ISRN Oncology* 2013 (2013); Xuan, C., et al. "Microbial dysbiosis is associated with human breast cancer." *PLoS One* 9, no. 1 (2014): e83744; Sheflin, A. M., et al. "Cancer-promoting effects of microbial dysbiosis." *Current Oncology Reports* 16, no. 10 (2014): 1-9; Kwa, M., et al. "The intestinal microbiome and estrogen receptor— positive female breast cancer." *Journal of the National Cancer Institute* 108, no. 8 (2016):

djw029; Plottel, C. S., et al. "Microbiome and malignancy." *Cell Host & Microbe* 10, no. 4 (2011): 324-35.

4. Cummings S. R., et al. "Prevention of breast cancer in postmenopausal women: Approaches to estimating and reducing risk." *Journal of the National Cancer Institute* 101, no. 6 (2009): 384-98.

5. Jolie, A. "My medical choice," *New York Times*, May 14, 2013, www.nytimes.com/2013/05/14 /opinion/my-medical-choice.html?_r=0.

6. Jolie, A. "Diary of a surgery," *New York Times*, March 24, 2015, www.nytimes.com/2015/03 /24/opinion/angelina-jolie-pitt-diary-of-a-surgery.html.

7. "The human genome project completion: Frequently asked questions." *National Human Genome Research Institute*, www.genome.gov/11006943; "Talking glossary of genetic terms," *National Human Genome Research Institute*, www.genome.gov/Glossary.

8. Shamovsky, I., et al. "New insights into the mechanism of heat shock response activation," *Cellular and Molecular Life Sciences* 65, no. 6 (2008): 855-61; Miozzo, F., et al. "HSFs, stress sensors and sculptors of transcription compartments and epigenetic landscapes," *Journal of Molecular Biology* 427, no. 24 (2015): 3793-3816; and Santoro, M. G. "Heat shock factors and the control of the stress response," *Biochemical Pharmacology* 59, no. 1 (2000): 55-63.

9. Yamashita, H., et al. "A glucose-responsive transcription factor that regulates carbohydrate metabolism in the liver," *Proceedings of the National Academy of Sciences* 98, no. 16 (2001): 9116-21.

10. Osborne, C. K., et al. "Estrogen receptor: current understanding of its activation and modulation," *Cancer Research* 7, no. 12 (2001): 4338s-42s; Halachmi, S., et al. "Estrogen receptor-associated proteins: possible mediators of hormone-induced transcription," *Science* 264, no. 5164 (1994): 1455-58; and Marino, M., et al. "Estrogen signaling multiple pathways to impact gene transcription," *Current Genomics* 7, no. 8 (2006): 497-508.

11. Audergon, P., et al. "Restricted epigenetic inheritance."

12. Sun, C., et al. "Potential epigenetic mechanism in non-alcoholic fatty liver disease." *International Journal of Molecular Sciences* 16, no. 3 (2015): 5161-79.

13. Er, T. K., et al. "Targeted next-generation sequencing for molecular diagnosis of endometriosis-associated ovarian cancer." *Journal of Molecular Medicine* (2016): 1-13; Wiegand, K. C., et al. "ARID1A mutations in endometriosis-associated ovarian carcinomas." *New England Journal of Medicine* 363, no. 16 (2010): 1532-43; Ayhan, A., et al. "Loss of ARID1A expression is an early molecular event in tumor progression from ovarian endometriotic cyst to clear cell and endometrioid carcinoma." *International Journal of Gynecological Cancer: Official Journal of the International Gynecological Cancer Society* 22, no. 8 (2012): 1310; Takeda, T., et al. "ARID1A gene mutation in ovarian and endometrial cancers (Review)." *Oncology Reports* 35, no. 2 (2016): 607-13.

14. Cao-Lei, L., et al. "DNA methylation signatures triggered by prenatal maternal stress exposure to a natural disaster: Project ice storm." *PLoS One* 9, no. 9 (2014).

Chapter 4: Get to the Root

1. Volpato, S., et al. "Cardiovascular disease, interleukin-6, and risk of mortality in older women the women's health and aging study." *Circulation* 103, no. 7 (2001): 947-53; Harris, T. B., et al. "Associations of elevated interleukin-6 and C-reactive protein levels with mortality in the elderly." *American Journal of Medicine* 106, no. 5 (1999): 506-12; Ferrucci, L., et al. "Serum IL-6 level and the development of disability in older persons." *Journal of the American Geriatrics Society* 47, no. 6 (1999): 639-46.

2. Lin, H., et al. "Whole blood gene expression and interleukin-6 levels." *Genomics* 104, no. 6 (2014): 490-95.

3. Barron, E., et al. "Blood-borne biomarkers of mortality risk: systematic review of cohort studies." *PLoS One* 10, no. 6 (2015): e0127550.

4. www.nhlbi.nih.gov/health/educational/lose_wt/BMI/bmicalc.htm.

5. Curtis, B. M., et al. "Autonomic tone as a cardiovascular risk factor: the dangers of chronic fight or flight." *Mayo Clinic Proceedings*, 77, no. 1 (2002): 45-54; Thayer, J. F., et al. "The role of vagal function in the risk for cardiovascular disease and mortality." *Biological Psychology* 74, no. 2 (2007): 224-42.

6. Kleiger, R. E., et al. "Heart rate variability: measurement and clinical utility." *Annals of Noninvasive Electrocardiology* 10, no. 1 (2005): 88–101; Dekker, J. M., et al. "Low heart rate variability in a 2-minute rhythm strip predicts risk of coronary heart disease and mortality from several causes The ARIC Study." *Circulation* 102, no. 11 (2000): 1239–44; Galinier, M. A., et al. "Depressed low frequency power of heart rate variability as an independent predictor of sudden death in chronic heart failure." *European Heart Journal* 21, no. 6 (2000): 475–82.

7. Buettner, D. "The island where people forget to die," *New York Times,* October 24, 2012, accessed August 17, 2015, www.nytimes.com/2012/10/28/magazine/the-island-where-people -forget-to-die.html?_r=1.

8. Panagiotakos, D. B., et al. "Sociodemographic and lifestyle statistics of oldest old people (> 80 years) living in Ikaria island: the Ikaria study." *Cardiology Research and Practice* (2011); Chrysohoou, C., et al. "Four-year (2009–2013) All cause and cardiovascular disease mortality and its determinants: The Ikaria study." *Journal of the American College of Cardiology* 63, no. 12_S (2014); Stefanadis, C. I., "Aging, genes and environment: lessons from the Ikaria study." *Hellenic Journal of Cardiology* 54, no. 3 (2013): 237–38; Trichopoulou, A., et al. "Anatomy of health effects of Mediterranean diet: Greek EPIC prospective cohort study." *British Medical Journal* 338 (2009).

9. Buettner, D. *The Blue Zones: 9 Lessons for Living Longer from the People Who've Lived the Longest,* 2nd ed. (Washington, DC: National Geographic, 2012).

10. Chilton, S. N., et al. "Inclusion of fermented foods in food guides around the world." *Nutrients* 7, no. 1 (2015): 390–404.

11. Timmers, S., et al. "Calorie restriction–like effects of 30 days of resveratrol supplementation on energy metabolism and metabolic profile in obese humans." *Cell Metabolism* 14, no. 5 (2011): 612–22; Morselli, E., et al. "Caloric restriction and resveratrol promote longevity through the Sirtuin-1-dependent induction of autophagy." *Cell Death and Disease* 1, no. 1 (2010): e10; Baur, J. A., et al. "Resveratrol improves health and survival of mice on a high-calorie diet." *Nature* 444, no. 7117 (2006): 337–42.

12. Pérez-Rubio, K. G., et al. "Effect of berberine administration on metabolic syndrome, insulin sensitivity, and insulin secretion." *Metabolic Syndrome and Related Disorders* 11, no. 5 (2013): 366–69; Pirillo, A., et al. "Berberine, a plant alkaloid with lipid-and glucose-lowering properties: From in vitro evidence to clinical studies." *Atherosclerosis* 243, no. 2 (2015): 449–61; Pang, B., et al. "Application of berberine on treating type 2 diabetes mellitus." *International Journal of Endocrinology* 2015 (2015).

13. Yarla, N. S., et al. "Targeting arachidonic acid pathway by natural products for cancer prevention and therapy." *Seminars in Cancer Biology* (2016); Zarei, A., et al. "A quick overview on some aspects of endocrinological and therapeutic effects of Berberis vulgaris L." *Avicenna Journal of Phytomedicine* 5, no. 6 (2015): 485; Caliceti, C., et al. "Potential benefits of berberine in the management of perimenopausal syndrome." *Oxidative Medicine and Cellular Longevity* (2015); Yang, J., et al. "Berberine improves insulin sensitivity by inhibiting fat store and adjusting adipokines profile in human preadipocytes and metabolic syndrome patients," *Evidence-Based Complementary and Alternative Medicine* 2012 (2012); and Hu, Y., et al. "Lipid-lowering effect of berberine in human subjects and rats," *Phytomedicine* 19, no. 10 (2012): 861–67.

14. Guo, Y., et al. "Repeated administration of berberine inhibits cytochromes P450 in humans," *European Journal of Clinical Pharmacology* 68, no. 2 (2012): 213–17.

Chapter 5: Feed—Week 1

1. Clark, M. "Still blazing trails," *New York Times,* August 4, 2014, accessed October 7, 2015, www.nytimes.com/2014/08/06/dining/still-blazing-trails.html.

2. Fussell, B. "Earning her food," *New York Times,* March 26, 2010, accessed September 20, 2015, www.nytimes.com/2010/03/28/magazine/28lives-t.html?_r=1.

3. Bergen, H. R., et al. "Myostatin as a mediator of sarcopenia versus homeostatic regulator of muscle mass: Insights using a new mass spectrometry-based assay." *Skeletal Muscle* 5, no. 1 (2015): 1.

4. Willer, C. J., et al. "Six new loci associated with body mass index highlight a neuronal

influence on body weight regulation." *Nature Genetics* 41, no. 1 (2009): 25–34; Wang, J., et al. "Study of eight GWAS-identified common variants for association with obesity-related indices in Chinese children at puberty." *International Journal of Obesity* 36, no. 4 (2012): 542–47; Speakman, J. R. "Functional analysis of seven genes linked to body mass index and adiposity by genome-wide association studies: a review." *Human Heredity* 75, nos. 2–4 (2013): 57–79; Fawcett, K. A., et al. "The genetics of obesity: FTO leads the way." *Trends in Genetics* 26, no. 6 (2010): 266–74.

5. Donaldson, C. M., et al. "Glycemic index and endurance performance." *International Journal of Sport Nutrition and Exercise Metabolism* 20, no. 2 (2010): 154–65; Bornet, F. R. J., et al. "Glycaemic response to foods: impact on satiety and long-term weight regulation." *Appetite* 49, no. 3 (2007): 535–53; Philippou, E., et al. "The influence of glycemic index on cognitive functioning: a systematic review of the evidence." *Advances in Nutrition: An International Review Journal* 5, no. 2 (2014): 119–30; Vranešić Bender, D., et al. "Nutritional and behavioral modification therapies of obesity: facts and fiction." *Digestive Diseases* 30, no. 2 (2012): 163–67; Mediano, M. F. F., et al. "Insulin Resistance Predicts the Effectiveness of Different Glycemic Index Diets on Weight Loss in Non-Obese Women." *Obesity Facts* 5, no. 5 (2012): 641–47; Sichieri, R., et al. "An 18-mo randomized trial of a low-glycemic-index diet and weight change in Brazilian women." *American Journal of Clinical Nutrition* 86, no. 3 (2007): 707–13.

6. Martins, M. L., et al. "Incidence of microflora and of ochratoxin A in green coffee beans (Coffea arabica)." *Food Additives and Contaminants* 20, no. 12 (2003): 1127–31; Studer-Rohr, I., et al. "The occurrence of ochratoxin A in coffee." *Food and Chemical Toxicology* 33, no. 5 (1995): 341–55.

7. Frankenfeld, C.L., et al. "High-intensity sweetener consumption and gut microbiome content and predicted gene function in a cross-sectional study of adults in the United States." *Annals of Epidemiology* 25, no. 10 (2015): 736–42; Burke, M. V., et al. "Physiological mechanisms by which non-nutritive sweeteners may impact body weight and metabolism." *Physiology and Behavior* (2015); Pepino, M. Y., "Metabolic effects of non-nutritive sweeteners." *Physiology and Behavior* (2015).

8. Chen, W. Y., et al. "Moderate alcohol consumption during adult life, drinking patterns, and breast cancer risk." *JAMA* 306, no. 17 (2011): 1884–90.

9. Iwai, K., et al. "Identification of food-derived collagen peptides in human blood after oral ingestion of gelatin hydrolysates." *Journal of Agricultural and Food Chemistry* 53, no. 16 (2005): 6531–36.

10. Choonpicharn, S., et al. "Antioxidant and antihypertensive activity of gelatin hydrolysate from Nile tilapia skin." *Journal of Food Science and Technology* 52, no. 5 (2014): 3134–39; Ao, J., et al. "Amino acid composition and antioxidant activities of hydrolysates and peptide fractions from porcine collagen." *Food Science and Technology International* 18, no. 5 (2012): 425–34.

11. Choonpicharn et al. "Antioxidant and antihypertensive activity"; Ngo, D. H., et al. "Angiotensin-I converting enzyme inhibitory peptides from antihypertensive skate (Okamejei kenojei) skin gelatin hydrolysate in spontaneously hypertensive rats." *Food Chemistry* 174 (2015): 37–43.

12. Leem, K. H., et al. "Porcine skin gelatin hydrolysate promotes longitudinal bone growth in adolescent rats." *Journal of Medicinal Food* 16, no. 5 (2013): 447–53.

13. Costanzo, S., et al. "Wine, beer or spirit drinking in relation to fatal and non-fatal cardiovascular events: a meta-analysis." *European Journal of Epidemiology* 26, no. 11 (2011): 833–50.

14. Park, K., et al. "Acute and subacute toxicity of copper sulfate pentahydrate (CuSO45 H2O) in the guppy (Poecilia reticulata)." *Journal of Veterinary Medical Science* 71, no. 3 (2009): 333–36; Hébert, C. D., et al. "Subchronic toxicity of cupric sulfate administered in drinking water and feed to rats and mice." *Fundamental and Applied Toxicology* 21, no. 4 (1993): 461–75; Sinković, A., et al. "Severe acute copper sulphate poisoning: a case report." *Archives of Industrial Hygiene and Toxicology* (2008): 31–35.

15. "All 48 fruits and vegetables with a pesticide residue data," *Environmental Working Group*, accessed June 15, 2015, www.ewg.org/foodnews/list.php.

16. Costanzo et al. "Wine, beer or spirit drinking in relation"; Streppel, M. T., et al. "Long-term wine consumption is related to cardiovascular mortality and life expectancy independently of moderate alcohol intake: the Zutphen Study." *Journal of Epidemiology and Community Health* 63, no. 7 (2009): 534–40.

17. Chen et al. "Moderate alcohol consumption"; Strumylaite, L., et al. "The association of low-to-moderate alcohol consumption with breast cancer subtypes defined by hormone receptor status." *PloS One* 10, no. 12 (2015): e0144680; Williams, L. A., et al. "Alcohol intake and invasive breast cancer risk by molecular subtype and race in the Carolina Breast Cancer Study." *Cancer Causes and Control* 27, no. 2 (2016): 259–69; Cao, Y., et al. "Light to moderate intake of alcohol, drinking patterns, and risk of cancer: results from two prospective US cohort studies." *British Medical Journal* (2015): h4238.

18. Goldberg, D. M., et al. "A global survey of trans-resveratrol concentrations in commercial wines." *American Journal of Enology and Viticulture* 46, no. 2 (1995): 159–65; Crandall, J. P., et al. "Pilot study of resveratrol in older adults with impaired glucose tolerance." *Journals of Gerontology Series A: Biological Sciences and Medical Sciences* (2012): glr235; Zamora-Ros, R., et al. "High urinary levels of resveratrol metabolites are associated with a reduction in the prevalence of cardiovascular risk factors in high-risk patients." *Pharmacological Research* 65, no. 6 (2012): 615–20; Brasnyó, P., et al. "Resveratrol improves insulin sensitivity, reduces oxidative stress and activates the Akt pathway in type 2 diabetic patients." *British Journal of Nutrition* 106, no. 3 (2011): 383–89; Marchal, J., et al. "Resveratrol in mammals: effects on aging biomarkers, age-related diseases, and life span." *Annals of the New York Academy of Sciences* 1290, no. 1 (2013): 67–73.

19. Semba, R. D., et al. "Resveratrol levels and all-cause mortality in older community-dwelling adults." *JAMA Internal Medicine* 174, no. 7 (2014): 1077–84; Yoshino, J., et al. "Resveratrol supplementation does not improve metabolic function in non-obese women with normal glucose tolerance." *Cell Metabolism* 16 (2012): 658–64; Bitterman, J. L., et al. "Metabolic effects of resveratrol: addressing the controversies." *Cellular and Molecular Life Sciences* 72, no. 8 (2015): 1473–88.

20. Timmers, S., et al. "Calorie restriction-like effects of 30 days of resveratrol supplementation on energy metabolism and metabolic profile in obese humans." *Cell Metabolism* 14, no. 5 (2011): 612–22; Morselli, E., et al. "Caloric restriction and resveratrol promote longevity through the sirtuin-1-dependent induction of autophagy." *Cell Death and Disease* 1, no. 1 (2010): e10; Baur, J. A., et al. "Resveratrol improves health and survival of mice on a high-calorie diet." *Nature* 444, no. 7117 (2006): 337–42.

21. Friedlander, B. "New York red wines show higher levels of resveratrol, a Cornell University study finds." *Cornell Chronicle,* February 2, 1998, accessed September 1, 2015. www.news .cornell.edu/stories/1998/02/ny-red-wines-show-more-resveratrol.

22. Paganini-Hill, A., et al. "Dental health behaviors, dentition, and mortality in the elderly: the leisure world cohort study." *Journal of Aging Research* (2011).

23. Olsen, I. "Update on bacteraemia related to dental procedures." *Transfusion and Apheresis Science* 39, no. 2 (2008): 173–78.

24. Akaji, E. A., et al. "Halitosis: a review of the literature on its prevalence, impact and control." *Oral Health and Preventative Dentistry* 12 (2014): 297–304.

25. Desvarieux, M., et al. "Periodontal microbiota and carotid intima-media thickness the oral infections and vascular disease epidemiology study (INVEST)." *Circulation* 111, no. 5 (2005): 576–82.

26. Yaacob, M., et al. "Powered versus manual toothbrushing for oral health." *Cochrane Database of Systematic Reviews* 6 (2014).

27. Desvarieux, M., et al. "Gender differences in the relationship between periodontal disease, tooth loss, and atherosclerosis." *Stroke* 35, no. 9 (2004): 2029–35; Wu, T., et al. "Periodontal disease and risk of cerebrovascular disease: the first national health and nutrition examination survey and its follow-up study." *Archives of Internal Medicine* 160, no. 18 (2000): 2749–55.

28. Peedikayil, F. C., et al. "Effect of coconut oil in plaque related gingivitis-A preliminary report." *Nigerian Medical Journal: Journal of the Nigeria Medical Association* 56, no. 2 (2015): 143; Asokan, S., et al. "Effect of oil pulling on plaque induced gingivitis: A randomized,

controlled, triple-blind study." *Indian Journal of Dental Research* 20, no. 1 (2009): 47; Roldan, S., et al. "Biofilms and the tongue: therapeutical approaches for the control of halitosis." *Clinical Oral Investigations* 7, no. 4 (2003): 189–97; Asokan, S., et al. "Effect of oil pulling on Streptococcus mutans count in plaque and saliva using Dentocult SM Strip mutans test: A randomized, controlled, triple-blind study." *Journal of Indian Society of Pedodontics and Preventive Dentistry* 26, no. 1 (2008): 12.

29. Sambunjak, D., et al. "Flossing for the management of periodontal diseases and dental caries in adults." *Cochrane Database of Systematic Reviews* 12 (2011).

Chapter 6: Sleep—Week 2

1. He, Y., et al. "The transcriptional repressor DEC2 regulates sleep length in mammals." *Science* 325, no. 5942 (2009): 866–70.

2. Gooley, J. J. "Circadian regulation of lipid metabolism." *The Proceedings of the Nutrition Society* (2016): 1–11; Gooley, J. J., et al. "Diurnal regulation of lipid metabolism and applications of circadian lipidomics." *Journal of Genetics and Genomics* 41, no. 5 (2014): 231–50; Horne, J. "The end of sleep: 'sleep debt' versus biological adaptation of human sleep to waking needs." *Biological Psychology* 87, no. 1 (2011): 1–14; Jackson, M. L., et al. "Cognitive components of simulated driving performance: sleep loss effects and predictors." *Accident Analysis & Prevention* 50 (2013): 438–44; McGrath, E., et al. "Sleep to lower elevated blood pressure: A randomized controlled trial (Slept)." *Journal of Hypertension* 34 (2016): e48; Wehr, T. A. "The durations of human melatonin secretion and sleep respond to changes in daylength (photoperiod)." *Journal of Clinical Endocrinology & Metabolism* 73, no. 6 (1991): 1276–80; Weintraub, K. "Ask well: Catching up on lost sleep," *New York Times,* July 24, 2015, accessed October 22, 2015, http://well.blogs.nytimes.com/2015/07/24/ask-well-catching-up-on-lost-sleep/?_r=0.

3. Archer, S. N., et al. "How sleep and wakefulness influence circadian rhythmicity: effects of insufficient and mistimed sleep on the animal and human transcriptome." *Journal of Sleep Research* 24, no. 5 (2015): 476–93.

4. Archer, S. N., et al. "Mistimed sleep disrupts circadian regulation of the human transcriptome." *Proceedings of the National Academy of Sciences* 111, no. 6 (2014): E682–91.

5. Archer et al. "How sleep and wakefulness influence circadian rhythmicity."

6. Tworoger, S. S., et al. "The association of self-reported sleep duration, difficulty sleeping, and snoring with cognitive function in older women." *Alzheimer Disease and Associated Disorders* 20, no. 1 (2006): 41–48.

7. Ferrie, J. E., et al. "Change in sleep duration and cognitive function: findings from the Whitehall II Study." *Sleep* 34, no. 5 (2011): 565.

8. Horne, J. "The end of sleep: 'sleep debt' versus biological adaptation of human sleep to waking needs." *Biological Psychology* 87, no. 1 (2011): 1–14.

9. Panagiotakos, D. B., et al. "Sociodemographic and lifestyle statistics of oldest old people (> 80 years) living in Ikaria island: the Ikaria study." *Cardiology Research and Practice* 2011 (2011).

10. Spiegel, K., et al. "Brief communication: Sleep curtailment in healthy young men is associated with decreased leptin levels, elevated ghrelin levels, and increased hunger and appetite." *Annals of Internal Medicine* 141, no. 11 (2004): 846–50; Taheri, S., et al. "Short sleep duration is associated with reduced leptin, elevated ghrelin, and increased body mass index." *PLoS Medicine* 1, no. 3 (2004): 210; Nedeltcheva, A. V., et al. "Sleep curtailment is accompanied by increased intake of calories from snacks." *American Journal of Clinical Nutrition* 89, no. 1 (2009): 126–33; Hart, C. N., et al. "Changes in children's sleep duration on food intake, weight, and leptin." *Pediatrics* 132, no. 6 (2013): e1473–80; Kjeldsen, J. S., et al. "Short sleep duration and large variability in sleep duration are independently associated with dietary risk factors for obesity in Danish school children." *International Journal of Obesity* 38, no. 1 (2014): 32–39; Leger, D., et al. "The role of sleep in the regulation of body weight." *Molecular and Cellular Endocrinology* (2015); Capers, P. L., et al. "A systemic review and meta-analysis of randomized controlled trials of the impact of sleep duration on adiposity and components of energy balance." *Obesity Reviews* 16, no. 9 (2015): 771–82;

Broussard, J. L., et al. "Elevated ghrelin predicts food intake during experimental sleep restriction." *Obesity* (2015).

11. Kim, T. W., et al. "The impact of sleep and circadian disturbance on hormones and metabolism." *International Journal of Endocrinology* (2015).

12. Spira, A. P., et al. "Self-reported sleep and β-amyloid deposition in community-dwelling older adults." *JAMA Neurology* 70, no. 12 (2013): 1537–43; Lim, A.S.P., et al. "Modification of the relationship of the apolipoprotein E ε4 allele to the risk of Alzheimer disease and neurofibrillary tangle density by sleep." *JAMA Neurology* 70, no. 12 (2013): 1544–51.

13. Kripke, D. F., et al. "Hypnotics' association with mortality or cancer: a matched cohort study." *British Medical Journal Open* 2, no. 1 (2012): e000850; Kripke, D. F. "Mortality risk of hypnotics: strengths and limits of evidence." *Drug Safety* (2015): 1–15; Mallon, L., et al. "Is usage of hypnotics associated with mortality?" *Sleep Medicine* 10, no. 3 (2009): 279–86.

14. Huedo-Medina, T. B., et al. "Effectiveness of non-benzodiazepine hypnotics in treatment of adult insomnia: meta-analysis of data submitted to the Food and Drug Administration." *British Medical Journal* 345 (2012): e8343.

15. Cedernaes, J., et al. "Acute sleep loss induces tissue-specific epigenetic and transcriptional alterations to circadian clock genes in men." *Journal of Clinical Endocrinology and Metabolism* 100, no. 9 (2015): E1255–61.

16. Jackson, M. L., et al. "Cognitive components of simulated driving performance: sleep loss effects and predictors." *Accident Analysis and Prevention* 50 (2013): 438–44.

17. Rattue, G. "Night shift working 'A probable human carcinogen.'" *Medical News Today,* October 28, 2011, accessed February 2, 2016. www.medicalnewstoday.com/articles/236731.php; "IARC monographs programme finds cancer hazards associated with shiftwork, painting and firefighting," *International Agency for Research on Cancer* December 5, 2007, accessed February 2, 2016, www.iarc.fr/en/media-centre/pr/2007/pr180.html; "Shiftwork," *IARC Monographs on the Evaluation of Carcinogenic Risks to Humans* 98 (2010), http://monographs.iarc.fr/ENG/Monographs/vol98/mono98-8.pdf.

18. Becker-Krail, D., et al. "Implications of circadian rhythm and stress in addiction vulnerability." *F1000Research* 5 (2016).

19. Roehrs, T., et al. "Caffeine: sleep and daytime sleepiness." *Sleep Medicine Reviews* 12, no. 2 (2008): 153–62.

20. Filipski, E., et al. "Effects of chronic jet lag on tumor progression in mice." *Cancer Research* 64, no. 21 (2004): 7879–85.

21. Wirz-Justice, A., et al. "Circadian disruption and psychiatric disorders: the importance of entrainment." *Sleep Medicine Clinics* 4, no. 2 (2009): 273–84; Davies, G., et al. "A systematic review of the nature and correlates of sleep disturbance in early psychosis." *Sleep Medicine Reviews* (2016).

22. Cajochen, C., et al. "Evening exposure to a light-emitting-diode (LED) backlit computer screen affects circadian physiology and cognitive performance." *Journal of Applied Physiology* 110, no. 5 (2011): 1432–38; Gooley, J. J., et al. "Exposure to room light before bedtime suppresses melatonin onset and shortens melatonin duration in humans." *Journal of Clinical Endocrinology and Metabolism* 96, no. 3 (2010): E463–72; Vinogradova, I. A., et al. "Circadian disruption induced by light-at-night accelerates aging and promotes tumorigenesis in rats." *Aging* 1, no. 10 (2009): 855.

23. Mallis, M. M., et al. "Circadian rhythms, sleep, and performance in space." *Aviation, Space, and Environmental Medicine* 76, no. Supplement 1 (2005): B94–B107.

24. Altevogt, B. M., et al., eds. *Sleep Disorders and Sleep Deprivation: An Unmet Public Health Problem* (Washington, DC: National Academies Press, 2006).

25. Copinschi, G. "Metabolic and endocrine effects of sleep deprivation." *Essential Psychopharmacology* 6, no. 6 (2004): 341–47.

26. Cohen, S., et al. "Sleep habits and susceptibility to the common cold." *Archives of Internal Medicine* 169, no. 1 (2009): 62–67; Opp, M. R. "Sleep and psychoneuroimmunology." *Immunology and Allergy Clinics of North America* 29, no. 2 (2009): 295–307; Krueger, J. M., et al. "Sleep, microbes and cytokines." *Neuroimmunomodulation* 1, no. 2 (1994): 100–109.

27. Hill, S. M., et al. "Melatonin: an inhibitor of breast cancer." *Endocrine-Related Cancer* (2015): ERC-15.

28. Stevens, R. G., et al. "Breast cancer and circadian disruption from electric lighting in the modern world." *CA: A Cancer Journal for Clinicians* 64, no. 3 (2014): 207–18; Hansen, J., et al. "Case–control study of shift-work and breast cancer risk in Danish nurses: impact of shift systems." *European Journal of Cancer* 48, no. 11 (2012): 1722–29; Knutsson, A., et al. "Breast cancer among shift workers: results of the WOLF longitudinal cohort study." *Scandinavian Journal of Work, Environment and Health* 39, no. 2 (2013): 170–77; Rabstein, S., et al. "Night work and breast cancer estrogen receptor status-results from the German GENICA study." *Scandinavian Journal of Work, Environment and Health* 39, no. 5 (2013): 448; Megdal, S. P., et al. "Night work and breast cancer risk: a systematic review and meta-analysis." *European Journal of Cancer* 41, no. 13 (2005): 2023–32.

29. Kamdar, B. B., et al. "Night-shift work and risk of breast cancer: a systematic review and meta-analysis." *Breast Cancer Research and Treatment* 138, no. 1 (2013): 291–301.

30. Maltese, F., et al. "Night shift decreases cognitive performance of ICU physicians." *Intensive Care Medicine* (2015): 1–8.

31. Bøggild, H., et al. "Shift work, risk factors and cardiovascular disease." *Scandinavian Journal of Work, Environment & Health* (1999): 85–99; Kawachi, I., et al. "Prospective study of shift work and risk of coronary heart disease in women." Circulation 92, no. 11 (1995): 3178–82; Vetter, C., et al. "Association Between Rotating Night Shift Work and Risk of Coronary Heart Disease Among Women." *Journal of the American Medical Association* 315, no. 16 (2016): 1726–34; Wang, A., et al. "Shift work and 20-year incidence of acute myocardial infarction: results from the Kuopio Ischemic Heart Disease Risk Factor Study." *Occupational and Environmental Medicine* (2016): oemed-2015.

32. Brown, D. L., et al. "Rotating night shift work and the risk of ischemic stroke." *American Journal of Epidemiology* (2009): kwp056.

33. Knutsson, A., et al. "Shift work and diabetes-A systematic review." *Chronobiology International* 31, no. 10 (2014): 1146–51; Pan, A., et al. "Rotating night shift work and risk of type 2 diabetes: two prospective cohort studies in women." *PLoS Medicine* 8, no. 12 (2011): 1660.

34. Bhatti, P., et al. "Nightshift work and risk of ovarian cancer." *Occupational and Environmental Medicine* 70, no. 4 (2013): 231–37; Hammer, G. P., et al. "Shift work and prostate cancer incidence in industrial workers: A historical cohort study in a German chemical company." *Deutsches Ärzteblatt International* 112, no. 27–28 (2015): 463; Hansen, J., et al. "Nested case-control study of night shift work and breast cancer risk among women in the Danish military." *Occupational and Environmental Medicine* 69, no. 8 (2012): 551–56; Heikkila, K., et al. "Long working hours and cancer risk: a multi-cohort study." *British Journal of Cancer* 114, no. 7 (2016): 813–18; Lin, X., "Night-shift work increases morbidity of breast cancer and all-cause mortality: a meta-analysis of 16 prospective cohort studies." *Sleep Medicine* 16, no. 11 (2015): 1381–87; Papantoniou, K., et al. "Increased and mistimed sex hormone production in night shift workers." *Cancer Epidemiology Biomarkers & Prevention* 24, no. 5 (2015): 854–63; Rao, D., et al. "Does night-shift work increase the risk of prostate cancer? a systematic review and meta-analysis." *OncoTargets and Therapy* 8 (2015): 2817; Reszka, E., et al. "Circadian genes in breast cancer." *Advances in Clinical Chemistry* (2016).

35. Hansen et al. "Case–control study of shift-work and breast cancer risk."

36. Phipps, A. I., et al. "Sleep duration and quality may impact cancer survival rate." *Sleep* 38 (2015).

37. Guarnieri, B., et al. "Sleep and cognitive decline: A strong bidirectional relationship. It is time for specific recommendations on routine assessment and the management of sleep disorders in patients with mild cognitive impairment and dementia." *European Neurology* 74, nos. 1–2 (2015): 43–48.

38. Blackwell, T., et al. "Poor sleep is associated with impaired cognitive function in older women: the study of osteoporotic fractures." *Journals of Gerontology Series A: Biological Sciences and Medical Sciences* 61, no. 4 (2006): 405–10.

39. Mander, B. A., et al. "Beta-amyloid disrupts human NREM slow waves and related hippocampus-dependent memory consolidation." *Nature Neuroscience* (2015).

40. Musiek, E. S., et al. "Sleep, circadian rhythms, and the pathogenesis of Alzheimer Disease." *Experimental and Molecular Medicine* 47, no. 3 (2015): e148.

41. Adan, A., et al. "Gender differences in morningness–eveningness preference." *Chronobiology International* 19, no. 4 (2002): 709–20.

42. Duffy, J. F., et al. "Sex difference in the near-24-hour intrinsic period of the human circadian timing system." *Proceedings of the National Academy of Sciences* 108, no. Suppl. 3 (2011): 15602–08; Lim, A. S. P., et al. "Sex difference in daily rhythms of clock gene expression in the aged human cerebral cortex." *Journal of Biological Rhythms* 28, no. 2 (2013): 117–29.

43. Roenneberg, T., et al "Epidemiology of the human circadian clock." *Sleep Medicine Reviews* 11, no. 6 (2007): 429–38.

44. Gominak, S. C., et al. "The world epidemic of sleep disorders is linked to vitamin D deficiency." *Medical Hypotheses* 79, no. 2 (2012): 132–35.

45. Gray, M. G., et al. "Multiple integrated complementary healing approaches: Energetics and light for bone." *Medical Hypotheses* 86 (2016): 18–29.

46. Shiue, I. "Low vitamin D levels in adults with longer time to fall asleep: US NHANES, 2005–2006." *International Journal of Cardiology* 41 (2013): 20–21.

47. Grandner, M. A., et al. "Relationships among dietary nutrients and subjective sleep, objective sleep, and napping in women." *Sleep Medicine* 11, no. 2 (2010): 180–84.

48. Massa, J., et al. "Vitamin D and actigraphic sleep outcomes in older community-dwelling men: the MrOS sleep study." *Sleep* 38, no. 2 (2014): 251–57.

49. Beydoun, M. A., et al. "Serum nutritional biomarkers and their associations with sleep among US adults in recent national surveys." *PloS One* 9, no. 8 (2014): e103490; Grandner, M. A., et al. "Sleep symptoms associated with intake of specific dietary nutrients." *Journal of Sleep Research* 23, no. 1 (2014): 22–34.

50. Shipton, E. A., et al. "Vitamin D and pain: Vitamin D and its role in the aetiology and maintenance of chronic pain states and associated comorbidities." *Pain Research and Treatment* 2015 (2015).

51. Balaban, H., et al. "Serum 25-hydroxyvitamin D levels in restless legs syndrome patients." *Sleep Medicine* 13, no. 7 (2012): 953–57; Wali, S. et al. "The effect of vitamin D supplements on the severity of restless legs syndrome." *Sleep and Breathing* 19, no. 2 (2015): 579–83; Gupta, R., et al. "Restless legs syndrome and pregnancy: prevalence, possible pathophysiological mechanisms and treatment." *Acta Neurologica Scandinavica* (2015).

52. Beydoun et al. "Serum nutritional biomarkers."

53. Xie, L., et al. "Sleep drives metabolite clearance from the adult brain." *Science* 342, no. 6156 (2013): 373–77; Jessen, N.A., et al. "The glymphatic system: A beginner's guide." *Neurochemical Research* (2015): 1–17; Tarasoff-Conway, J. M., et al. "Clearance systems in the brain—implications for Alzheimer disease." *Nature Reviews Neurology* 11, no. 8 (2015): 457–70; Mendelsohn, A. R., et al. "Sleep facilitates clearance of metabolites from the brain: glymphatic function in aging and neurodegenerative diseases." *Rejuvenation Research* 16, no. 6 (2013): 518–23.

54. Lee, H., et al. "The effect of body posture on brain glymphatic transport." *Journal of Neuroscience* 35, no. 31 (2015): 11034–44.

55. Wang, T. J., et al. "Common genetic determinants of vitamin D insufficiency: a genome-wide association study." *Lancet* 376, no. 9736 (2010): 180–88.

56. Ross, A. C., et al., eds. *Dietary reference intakes for calcium and vitamin D* (Washington, DC: National Academies Press, 2010).

57. Crowley, S. J., et al. "Increased sensitivity of the circadian system to light in early/mid-puberty." *Journal of Clinical Endocrinology and Metabolism* 100, no. 11 (2015): 4067–73.

58. Tamakoshi, A., et al. "Self-reported sleep duration as a predictor of all-cause mortality: results from the JACC study, Japan." *Sleep* 27, no. 1 (2004): 51–54; Hublin, C., et al. "Sleep and mortality: a population-based 22-year follow-up study." *Sleep* 30, no. 10 (2007): 1245; Gallicchio, L., et al. "Sleep duration and mortality: a systematic review and meta analysis." *Journal of Sleep Research* 18, no. 2 (2009): 148–58.

59. Youngstedt, S. D., et al. "Long sleep and mortality: rationale for sleep restriction." *Sleep Medicine Reviews* 8, no. 3 (2004): 159–74.

60. Sofer, S., et al. "Greater weight loss and hormonal changes after 6 months diet with carbohydrates eaten mostly at dinner." *Obesity* 19, no. 10 (2011): 2006–14; Sofer, S., et al. "Changes in daily leptin, ghrelin and adiponectin profiles following a diet with

carbohydrates eaten at dinner in obese subjects." *Nutrition, Metabolism and Cardiovascular Diseases* 23, no. 8 (2013): 744–50.

61. Richards, J., et al. "Higher serum vitamin D concentrations are associated with longer leukocyte telomere length in women." *American Journal of Clinical Nutrition* 86, no. 5 (2007): 1420–25; Liu, J. J., et al. "Plasma vitamin D biomarkers and leukocyte telomere length." *American Journal of Epidemiology* (2013): kws435.

62. Satlin, A., et al. "Bright light treatment of behavioral and sleep disturbances in patients with Alzheimer's disease." *American Journal of Psychiatry* 149 (1992): 1028–32; Mishima, K., et al. "Morning bright light therapy for sleep and behavior disorders in elderly patients with dementia." *Acta Psychiatrica Scandinavica* 89, no. 1 (1994): 1–7; Stewart, K. T., et al. "Light treatment for NASA shiftworkers." *Chronobiology International* 12, no. 2 (1995): 141–151; Mishima, K., et al. "Randomized, dim light controlled, crossover test of morning bright light therapy for rest-activity rhythm disorders in patients with vascular dementia and dementia of Alzheimer's type." *Chronobiology International* 15, no. 6 (1998): 647–54; Lyketsos, C. G., et al. "A randomized, controlled trial of bright light therapy for agitated behaviors in dementia patients residing in long-term care." *International Journal of Geriatric Psychiatry* 14, no. 7 (1999): 520–25; Yamadera, H., et al. "Effects of bright light on cognitive and sleep–wake (circadian) rhythm disturbances in Alzheimer-type dementia." *Psychiatry and Clinical Neurosciences* 54, no. 3 (2000): 352–53; Ancoli-Israel, S., et al. "Increased light exposure consolidates sleep and strengthens circadian rhythms in severe Alzheimer's disease patients." *Behavioral Sleep Medicine* 1, no. 1 (2003): 22–36; Fetveit, A., et al. "Bright light treatment improves sleep in institutionalised elderly—an open trial." *International Journal of Geriatric Psychiatry* 18, no. 6 (2003): 520–26.

63. Lockley, S. W., et al. "High sensitivity of the human circadian melatonin rhythm to resetting by short wavelength light." *Journal of Clinical Endocrinology and Metabolism* 88, no. 9 (2003): 4502–05; Sasseville, A., et al. "Wearing blue-blockers in the morning could improve sleep of workers on a permanent night schedule: a pilot study." *Chronobiology International* 26, no. 5 (2009): 913–25; Wood, B., et al. "Light level and duration of exposure determine the impact of self-luminous tablets on melatonin suppression." *Applied Ergonomics* 44, no. 2 (2013): 237–40; van der Lely, S., et al. "Blue blocker glasses as a countermeasure for alerting effects of evening light-emitting diode screen exposure in male teenagers." *Journal of Adolescent Health* 56, no. 1 (2015): 113–19.

64. Duffy, J. F., et al. "Sex difference in the near-24-hour intrinsic period of the human circadian timing system." *Proceedings of the National Academy of Sciences* 108, no. Supplement 3 (2011): 15602–8.

Chapter 7: Move—Week 3

1. "Health risks of physically strenuous work," *European Observatory of Working Life*, March 8, 2005, accessed November 2, 2015, www.eurofound.europa.eu/observatories/ eurwork/articles/working-conditions/health-risks-of-physically-strenuous-work; Künzler, G., et al. "Arme sterben früher: soziale Schicht, mortalität und rentenalterspolitik in der Schweiz." Vol. 11. *Caritas-Verlag* (2002); "National census of fatal occupational injuries in 2014," *Bureau of Labor Statistics, US Department of Labor*, September 17, 2015, accessed November 2, 2015, www.bls.gov/news.release/pdf/cfoi.pdf; Raley, D. "New NFL goal: A longer life," *Seattle Pi*, May 8, 2008, accessed November 2, 2015, www.seattlepi.com/news /article/New-NFL-goal-A-longer-life-1272886.php.

2. Matthews, C. E., et al. "Amount of time spent in sedentary behaviors in the United States, 2003–2004." *American Journal of Epidemiology* 167, no. 7 (2008): 875–81.

3. Patel, A. V., et al. "Leisure-time spent sitting and site-specific cancer incidence in a large US cohort." *Cancer Epidemiology Biomarkers and Prevention* 24, no. 9 (2015): 1350–59.

4. Shibata, A., et al. "Physical activity, television viewing time and 12-year changes in waist circumference." *Medicine and Science in Sports and Exercise* (2015); Chastin, S.F.M., et al. "Combined effects of time spent in physical activity, sedentary behaviors and sleep on obesity and cardio-metabolic health markers: A novel compositional data analysis approach." *PloS One* 10, no. 10 (2015): e0139984.

5. Chastin et al. "Combined effects of time spent in physical activity"; Lamb, M. J. E., et al. "Prospective associations between sedentary time, physical activity, fitness and cardiometabolic risk factors in people with type 2 diabetes." *Diabetologia* 59, no. 1 (2016): 110–20.
6. Wilmot, E. G., et al. "Sedentary time in adults and the association with diabetes, cardiovascular disease and death: systematic review and meta-analysis." *Diabetologia* 55 (2012): 2895–05; Warburton, D., et al. "Health benefits of physical activity: the evidence." *Canadian Medical Association Journal* 174, no. 6 (2006): 801–09.
7. Owen, N., et al. "Too much sitting: a novel and important predictor of chronic disease risk?" *British Journal of Sports Medicine* 43, no. 2 (2009): 81–83; Bauman, A., et al. "Leisure-time physical activity alone may not be a sufficient public health approach to prevent obesity—a focus on China." *Obesity Reviews* 9, no. s1 (2008): 119–26.
8. Chastin et al. "Combined effects of time spent in physical activity."
9. Samitz, G, et al. "Domains of physical activity and all-cause mortality: systematic review and dose–response meta-analysis of cohort studies." *International Journal of Epidemiology* 40, no. 5 (2011): 1382–1400; Hu, G., et al. "The effects of physical activity and body mass index on cardiovascular, cancer and all-cause mortality among 47 212 middle-aged Finnish men and women." *International Journal of Obesity* 29, no. 8 (2005): 894–902. Schnohr, P., et al. "Longevity in male and female joggers: the Copenhagen City Heart Study." *American Journal of Epidemiology* 177, no. 7 (2013): 683–89.
10. Oguma, Y., et al. "Physical activity and all-cause mortality in women: a review of the evidence." *British Journal of Sports Medicine* 36, no. 3 (2002): 162–72.
11. Gregg, E. W., et al. "Relationship of changes in physical activity and mortality among older women." *JAMA* 289, no. 18 (2003): 2379–86.
12. Stanford, K. I., et al. "Exercise effects on white adipose tissue: Beiging and metabolic adaptations." *Diabetes* 64, no. 7 (2015): 2361–68.
13. Levine, H. J. "Rest heart rate and life expectancy." *Journal of the American College of Cardiology* 30, no. 4 (1997): 1104–6.
14. "Physical activity guidelines," *Health.Gov*, accessed December 7, 2015, http://health.gov /paguidelines.
15. Schnohr, P., et al. "Dose of jogging and long-term mortality: the Copenhagen City Heart Study." *Journal of the American College of Cardiology* 65, no. 5 (2015): 411–19.
16. Lavie, C. J., et al. "Effects of running on chronic diseases and cardiovascular and all-cause mortality." *Mayo Clinic Proceedings* 90, no. 11 (2015): 1541–52.
17. Day, S. M., et al. "Cardiac risks associated with marathon running." *Sports Health: A Multidisciplinary Approach* 2, no. 4 (2010): 301–6; Kim, J. H., et al. "Cardiac arrest during long-distance running races." *New England Journal of Medicine* 366, no. 2 (2012): 130–40; Hart, L. "Marathon-related cardiac arrest." *Clinical Journal of Sport Medicine* 23, no. 5 (2013): 409–10.
18. Du, M., et al. "Physical activity, sedentary behavior, and leukocyte telomere length in women." *American Journal of Epidemiology* (2012): kwr330.
19. Ibid.
20. Ibid.; Krishna, B. H., et al. "Association of leukocyte telomere length with oxidative stress in yoga practitioners." *Journal of Clinical and Diagnostic Research: JCDR* 9, no. 3 (2015): CC01.
21. Martyn–St. James, M., et al. "Meta-analysis of walking for preservation of bone mineral density in postmenopausal women." *Bone* 43, no. 3 (2008): 521–31.
22. Zhao, R., et al. "The effects of differing resistance training modes on the preservation of bone mineral density in postmenopausal women: a meta-analysis." *Osteoporosis International* 26, no. 5 (2015): 1605–18.
23. Patel, N. K., et al. "The effects of yoga on physical functioning and health related quality of life in older adults: a systematic review and meta-analysis." *Journal of Alternative and Complementary Medicine* 18, no. 10 (2012): 902–17; Phoosuwan, M., et al. "The effects of weight bearing yoga training on the bone resorption markers of the postmenopausal women." *Chotmaihet Thangphaet [Journal of the Medical Association of Thailand]* 92 (2009): S102–8.
24. Melov, S. et al. "Resistance exercise reverses aging in human skeletal muscle." *PLoS One* 2, no. 5 (2007): e465.

25. Lee, J. A., et al. "Effects of yoga exercise on serum adiponectin and metabolic syndrome factors in obese postmenopausal women." *Menopause* 19, no. 3 (2012): 296–301.

26. Watson, K., et al. "MTOR and the health benefits of exercise." *Seminars in Cell and Developmental Biology*, no. 36, (2014): 130–39.

27. Markofski, M. M., et al. "Effect of age on basal muscle protein synthesis and mTORC1 signaling in a large cohort of young and older men and women." *Experimental Gerontology* 65 (2015): 1–7.

28. Rönn, T., et al. "A six months exercise intervention influences the genome-wide DNA methylation pattern in human adipose tissue." *PLoS Genetics* 9, no. 6 (2013): e1003572.

29. Vimaleswaran, K. S., et al. "Physical activity attenuates the body mass index–increasing influence of genetic variation in the FTO gene." *American Journal of Clinical Nutrition* 90, no. 2 (2009): 425–28; Kilpeläinen, T. O., et al. "Physical activity attenuates the influence of FTO variants on obesity risk: a meta-analysis of 218,166 adults and 19,268 children." *PLoS Medicine* 8, no. 11 (2011): e1001116; Shengxu, L., et al. "Cumulative effects and predictive value of common obesity-susceptibility variants identified by genome-wide association studies." *American Journal of Clinical Nutrition* 91, no. 1 (2010): 184–90.

30. Kilpeläinen, T. O., et al. "Genome-wide meta-analysis uncovers novel loci influencing circulating leptin levels." *Nature Communications* 7 (2016).

31. Dupuis, J., et al. "New genetic loci implicated in fasting glucose homeostasis and their impact on type 2 diabetes risk." *Nature Genetics* 42, no. 2 (2010): 105–16; Strawbridge, R. J., et al. "Genome-wide association identifies nine common variants associated with fasting proinsulin levels and provides new insights into the pathophysiology of type 2 diabetes." *Diabetes* 60, no. 10 (2011): 2624–34; Teran-Garcia, M., et al. "Hepatic lipase gene variant—514C>T is associated with lipoprotein and insulin sensitivity response to regular exercise." *Diabetes* 54, no. 7 (2005): 2251–55.

32. Ahmad, T., et al. "Physical activity modifies the effect of LPL, LIPC, and CETP polymorphisms on HDL-C levels and the risk of myocardial infarction in women of European ancestry." *Circulation: Cardiovascular Genetics* 4, no. 1 (2011): 74–80.

33. Barres, R., et al. "Acute exercise remodels promoter methylation in human skeletal muscle." *Cell Metabolism* 15, no. 3 (2012): 405–11.

34. Hargreaves, M. "Exercise and Gene Expression." *Progress in Molecular Biology and Translational Science* 135 (2015): 457–69.

35. Bratman, G. N., et al. "The benefits of nature experience: Improved affect and cognition." *Landscape and Urban Planning* 138 (2015): 41–50.

36. Bratman, G. N., et al. "Nature experience reduces rumination and subgenual prefrontal cortex activation." *Proceedings of the National Academy of Sciences* 112, no. 28 (2015): 8567–72.

37. Song, C., et al. "Physiological and psychological responses of young males during springtime walks in urban parks." *Journal of Physiological Anthropology* 33, no. 8 (2014); Song, C., et al. "Physiological and psychological effects of walking on young males in urban parks in winter." *Journal of Physiological Anthropology* 32, no. 1 (2013): 18; Song, C., et al. "Effect of forest walking on autonomic nervous system activity in middle-aged hypertensive individuals: A pilot study." *International Journal of Environmental Research and Public Health* 12, no. 3 (2015): 2687–99.

38. Gammon, M. D., et al. "Recreational and occupational physical activities and risk of breast cancer." *Journal of the National Cancer Institute* 90, no. 2 (1998): 100–117.

39. Dorn, J., et al. "Lifetime physical activity and breast cancer risk in pre-and postmenopausal women." *Medicine and Science in Sports and Exercise* 35, no. 2 (2003): 278–85.

40. Patel, A. V., et al. "Recreational physical activity and risk of postmenopausal breast cancer in a large cohort of US women." *Cancer Causes and Control* 14, no. 6 (2003): 519–29.

41. Patel, A. V., et al. "Lifetime recreational exercise activity and risk of breast carcinoma in situ." *Cancer* 98, no. 10 (2003): 2161–69; Lu, Y., et al. "History of recreational physical activity and survival after breast cancer: the California Breast Cancer Survivorship Consortium." *American Journal of Epidemiology* (2015): kwu466; Warburton et al. "Health benefits of physical activity."

42. Lek, M., et al. "Analysis of protein-coding genetic variation in 60,706 humans." *bioRxiv* (2015): 030338.

43. Rankinen, T., et al. "Effect of endothelin 1 genotype on blood pressure is dependent on physical activity or fitness levels." *Hypertension* 50, no. 6 (2007): 1120–25; Li, T. C., et al. "Associations of EDNRA and EDN1 polymorphisms with carotid intima media thickness through interactions with gender, regular exercise, and obesity in subjects in Taiwan: Taichung Community Health Study (TCHS)." *BioMedicine* 5, no. 2 (2015).

44. Yaffe, K., et al. "A prospective study of physical activity and cognitive decline in elderly women: women who walk." *Archives of Internal Medicine* 161, no. 14 (2001): 1703–8.

45. Weuve, J., et al. "Physical activity, including walking, and cognitive function in older women." *JAMA* 292, no. 12 (2004): 1454–61.

46. Zulfiqar, U., et al. "Relation of high heart rate variability to healthy longevity." *American Journal of Cardiology* 105, no. 8 (2010): 1181–85.

47. Blomstrand, A., et al. "Effects of leisure time physical activity on well-being among women in a 32-year perspective." *Scandinavian Journal of Public Health* (2009).

48. Crane, J. D., et al. "Exercise-stimulated interleukin-15 is controlled by AMPK and regulates skin metabolism and aging." *Aging Cell* (2015).

49. Li, F., et al. "Tai chi and self-rated quality of sleep and daytime sleepiness in older adults: A randomized controlled trial." *Journal of the American Geriatrics Society* 52, no. 6 (2004): 892–900.

50. Irwin, M. R., et al. "Tai chi, cellular inflammation, and transcriptome dynamics in breast cancer survivors with insomnia: a randomized controlled trial." *Journal of the National Cancer Institute. Monographs* 2014, no. 50 (2014): 295–301.

51. Buman, M. P., et al. "Does nighttime exercise really disturb sleep? Results from the 2013 National Sleep Foundation Sleep in America poll." *Sleep Medicine* 15, no. 7 (2014): 755–61; Qian-Chun, Y., et al. "Impact of evening exercise on college students' sleep quality." *Zhonghua Yu Fang Yi Xue Za Zhi [Chinese Journal of Preventive Medicine]* 47, no. 6 (2013): 542–46; Brand, S., et al. "High self-perceived exercise exertion before bedtime is associated with greater objectively assessed sleep efficiency." *Sleep Medicine* 15, no. 9 (2014): 1031–36. Alley, J. R., et al. "Effects of resistance exercise timing on sleep architecture and nocturnal blood pressure." *Journal of Strength and Conditioning Research* 29, no. 5 (2015): 1378–85.

52. Yang, P. Y., et al. "Exercise training improves sleep quality in middle-aged and older adults with sleep problems: a systematic review." *Journal of Physiotherapy* 58, no. 3 (2012): 157–63.

53. Fiatarone, M. A., et al. "High-intensity strength training in nonagenarians: effects on skeletal muscle." *Journal of the American Medical Association* 263, no. 22 (1990): 3029–34.

54. 58 Poirier, P. "Exercise, heart rate variability, and longevity: the cocoon mystery?" *Circulation* 129 (2014): 2085–87; U.S. Department of Health and Human Services. Physical Activity Guidelines for Americans: Be Active, Healthy and Happy (Washington, DC: US Department of Health and Human Services, 2008); Myers, J., et al. "Exercise capacity and mortality among men referred for exercise testing." *New England Journal of Medicine* 346, no. 11 (2002): 793–801; Gulati, M., et al. "The prognostic value of a nomogram for exercise capacity in women." *New England Journal of Medicine* 353, no. 5 (2005): 468–75.

55. Reynolds, G. "Walk hard. Walk easy. Repeat," *New York Times,* February 19, 2015, accessed September 20, 2015, http://well.blogs.nytimes.com/2015/02/19/walk-hard-walk-easy-repeat/?_r=0; Masuki, S., et al. "The factors affecting adherence to a long-term interval walking training program in middle-aged and older people." *Journal of Applied Physiology* 118, no. 5 (2015): 595–603.

56. Esmarck, B., et al. "Timing of post exercise protein intake is important for muscle hypertrophy with resistance training in elderly humans." *Journal of Physiology* 535, no. 1 (2001): 301–11.

57. Buchheit, M., et al. "Parasympathetic reactivation after repeated sprint exercise." *American Journal of Physiology-Heart and Circulatory Physiology* 293, no. 1 (2007): H133-H141; Bishop, P. A., et al. "Recovery from training: a brief review: brief review." *Journal of Strength and Conditioning Research* 22, no. 3 (2008): 1015–24; Gisselman, A. et al. "Musculoskeletal overuse injuries and heart rate variability: Is there a link?" *Medical Hypotheses* 87 (2016):

1–7; Mayo, X., et al. "Exercise type affects cardiac vagal autonomic recovery after a resistance training session." *Journal of Strength and Conditioning Research / National Strength and Conditioning Association* (2016); Vernillo, G., et al. "Postexercise autonomic function after repeated-sprints training." *European Journal of Applied Physiology* 115, no. 11 (2015): 2445–55.

58. Gupta, S., et al. "Cardiorespiratory fitness and classification of risk of cardiovascular disease mortality." *Circulation* 123, no. 13 (2011): 1377–83; Berry, J. D., et al. "Lifetime risks for cardiovascular disease mortality by cardiorespiratory fitness levels measured at ages 45, 55, and 65 years in men: the Cooper Center Longitudinal Study." *Journal of the American College of Cardiology* 57, no. 15 (2011): 1604–10.

59. Waygood, E., et al. "Active travel by built environment and lifecycle stage: case study of Osaka metropolitan area." *International Journal of Environmental Research and Public Health* 12, no. 12 (2015): 15900–24; Oja, P., et al. "Health benefits of cycling: a systematic review." *Scandinavian Journal of Medicine and Science in Sports* 21, no. 4 (2011): 496–509; Chiu, M, et al. "Moving to a Highly Walkable Neighborhood and Incidence of Hypertension: A Propensity-score Matched Cohort Study." *Circulation* 132, no. Suppl. 3 (2015): A11545.

60. Martin, A., et al. "Does active commuting improve psychological wellbeing? Longitudinal evidence from eighteen waves of the British Household Panel Survey." *Preventive Medicine* 69 (2014): 296–303; Gatersleben, B., et al. "Affective appraisals of the daily commute comparing perceptions of drivers, cyclists, walkers, and users of public transport." *Environment and Behavior* 39, no. 3 (2007): 416–31.

Chapter 8: Release—Week 4

1. Krishna, B. H., et al. "Association of leukocyte telomere length with oxidative stress in yoga practitioners." *Journal of Clinical and Diagnostic Research: JCDR* 9, no. 3 (2015): CC01; Balasubramanian, S., et al. "Induction of salivary nerve growth factor by Yogic breathing: a randomized controlled trial." *International Psychogeriatrics* 27, no. 1 (2015): 168–70; Bower, J. E., et al. "Yoga reduces inflammatory signaling in fatigued breast cancer survivors: a randomized controlled trial." *Psychoneuroendocrinology* 43 (2014): 20–29; Qu, S., et al. "Rapid gene expression changes in peripheral blood lymphocytes upon practice of a comprehensive yoga program." *PloS One* 8, no. 4 (2013): e61910; Sharma, H., et al. "Sudarshan Kriya practitioners exhibit better antioxidant status and lower blood lactate levels." *Biological Psychology* 63, no. 3 (2003): 281–91.

2. Inanir, A., et al. "Clinical symptoms in fibromyalgia are associated to catechol-O-methyltransferase (COMT) gene Val158Met polymorphism." *Xenobiotica* 44, no. 10 (2014): 952–56.

3. Vossen, H., et al. "The genetic influence on the cortical processing of experimental pain and the moderating effect of pain status." *PLoS One* 5, no. 10 (2010): e13641; Nijs, J., et al. "Brain-derived neurotrophic factor as a driving force behind neuroplasticity in neuropathic and central sensitization pain: a new therapeutic target?" *Expert Opinion on Therapeutic Targets* 19, no. 4 (2015): 565–76.

4. Forrest, A. *Fierce Medicine* (San Francisco: HarperOne, 2012).

5. For more information on Forrest yoga, visit www.forrestyoga.com.

6. Kogan, L. "Oprah's new workout and what happened when we tried it)." *Oprah.com*, January 2016, accessed January 26, 2016, www.oprah.com/health/Oprahs-New-Workout-and-What-Happened-When-We-Tried-It.

7. Bougea, A. M., et al. "Effect of the emotional freedom technique on perceived stress, quality of life, and cortisol salivary levels in tension-type headache sufferers: a randomized controlled trial." *Explore: Journal of Science and Healing* 9, no. 2 (2013): 91–99.

8. Church, D., et al. "The effect of emotional freedom techniques on stress biochemistry: a randomized controlled trial." *Journal of Nervous and Mental Disease* 200, no. 10 (2012): 891–96.

9. Brattberg, G. "Self-administered EFT (Emotional Freedom Techniques) in individuals with fibromyalgia: a randomized trial." *Integrative Medicine* 7, no. 4 (2008): 30–35.

10. Boath, E., et al. "A narrative systematic review of the effectiveness of Emotional Freedom

Techniques (EFT)." *Staffordshire University, CPSI Monograph, Centre for Practice and Service Improvement* (2012).

11. Ibid.

12. Upledger, J. E., et al. *CranioSacral Therapy: What It Is, How It Works* (Berkeley: North Atlantic Books, 2008), 103.

13. Henschke, N., et al. "Stretching before or after exercise does not reduce delayed-onset muscle soreness." *British Journal of Sports Medicine* 45, no. 15 (2011): 1249–50.

14. Simic, L., N. et al. "Does pre-exercise static stretching inhibit maximal muscular performance? A meta-analytical review." *Scandinavian Journal of Medicine and Science in Sports* 23, no. 2 (2013): 131–48.

15. Cheatham, S. W., et al. "The effects of self-myofascial release using a foam roll or roller massager on joint range of motion, muscle recovery and performance: A systematic review." *International Journal of Sports Physical Therapy* 10, no. 6 (2015): 827.

16. Beardsley, C., et al. "Effects of self-myofascial release: A systematic review." *Journal of Bodywork and Movement Therapies* 19, no. 4 (2015): 747–58.

17. Chen, Y. H., et al. "Increased sliding of transverse abdominis during contraction after myofascial release in patients with chronic low back pain." *Manual Therapy* (2015).

18. Bleakley, C., et al. "Cold-water immersion for preventing and treating muscle soreness after exercise." *Cochrane Library* (2012); Costello, J. T., et al. "Whole-body cryotherapy (extreme cold air exposure) for preventing and treating muscle soreness after exercise in adults." *Cochrane Library* (2015).

19. Bleakley, C. M., et al. "What is the biochemical and physiological rationale for using Cold Water Immersion in Sports Recovery? A Systematic Review." *British Journal of Sports Medicine* (2009): bjsm-2009.

20. Van der Kolk, B. *The Body Keeps the Score* (New York: Viking, 2014), 356.

21. Ferris, T. "Relax like a pro: 5 steps to hacking your sleep," *The Tim Ferris Experiment*, January 27, 2008, accessed February 1, 2016, http://fourhourworkweek.com/2008/01/27 /relax-like-a-pro-5-steps-to-hacking-your-sleep.

22. Hamblin, J. "The benefits of being cold," *Atlantic*, January 2015, accessed February 1, 2016, www.theatlantic.com/magazine/archive/2015/01/does-global-warming-make-me -look-fat/383509.

23. Zhornitsky, S., et al. "Cannabidiol in humans—the quest for therapeutic targets," *Pharmaceuticals* 5, no. 5 (2012): 529–52; Welty, T. E., et al. "Cannabidiol: promise and pitfalls," *Epilepsy Currents* 14, no. 5 (2014): 250–52; and Fasinu, P. S., et al. "Current status and prospects for cannabidiol preparations as new therapeutic agents," *Pharmacotherapy: The Journal of Human Pharmacology and Drug Therapy* (2016).

Chapter 9: Expose—Week 5

1. Gore, A. C., et al. "Executive summary to EDC-2: The Endocrine Society's second scientific statement on endocrine-disrupting chemicals." *Endocrine Reviews* 36, no. 6 (2015): 593–602.

2. Christensen, K., et al. "Ageing populations: the challenges ahead." *Lancet* 374, no. 9696 (2009): 1196–1208.

3. Wild, C. P. "Complementing the genome with an 'exposome': the outstanding challenge of environmental exposure measurement in molecular epidemiology." *Cancer Epidemiology Biomarkers and Prevention* 14, no. 8 (2005): 1847–50.

4. Coughlin, S. S., et al. "The impact of the natural, social, built, and policy environments on breast cancer." *Journal of Environment and Health Sciences* 1, no. 3 (2015); Land, C. E. "Studies of cancer and radiation dose among atomic bomb survivors: the example of breast cancer." *JAMA* 274, no. 5 (1995): 402–7; Hancock, S. L., et al. "Breast cancer after treatment of Hodgkin's disease." *Journal of the National Cancer Institute* 85, no. 1 (1993): 25–31.

5. Inspired in part by the International Living Future Institute's red list of building materials that contain harmful substances. "Materials red list," *International Living Future Institute*, accessed February 6, 2016, http://living-future.org/topic/materials-red-list.

6. Kojima, H., et al. "In vitro endocrine disruption potential of organophosphate flame retardants via human nuclear receptors." *Toxicology* 314, no. 1 (2013): 76–83.

7. Liu, X., et al. "Endocrine disruption potentials of organophosphate flame retardants and related mechanisms in H295R and MVLN cell lines and in zebrafish." *Aquatic Toxicology* 114 (2012): 173–81.

8. Su, G., et al. "Rapid in vitro metabolism of the flame retardant triphenyl phosphate and effects on cytotoxicity and mRNA expression in chicken embryonic hepatocytes." *Environmental Science and Technology* 48, no. 22 (2014): 13511–19.

9. Mendelsohn, E., et al. "Nail polish as a source of exposure to triphenyl phosphate." *Environment International* 86 (2016): 45–51.

10. Calle, E. E., et al. "Organochlorines and breast cancer risk." *CA: A Cancer Journal for Clinicians* 52, no. 5 (2002): 301–9; Hertz-Picciotto, I., ed. *Breast Cancer and the Environment: A Life Course Approach* (National Academies Press, 2012); Ekenga, C. C., et al. "Breast cancer risk after occupational solvent exposure: The influence of timing and setting." *Cancer Research* 74, no. 11 (2014): 3076–83; Labrèche, F., et al. "Postmenopausal breast cancer and occupational exposures." *Occupational and Environmental Medicine* 67, no. 4 (2010): 263–69; Millikan, R., et al. "Dichlorodiphenyldichloroethene, polychlorinated biphenyls, and breast cancer among African-American and white women in North Carolina." *Cancer Epidemiology Biomarkers and Prevention* 9, no. 11 (2000): 1233–40; Krieger, N., et al. "Breast cancer and serum organochlorines: a prospective study among white, black, and Asian women." *Journal of the National Cancer Institute* 86, no. 8 (1994): 589–99.

11. Kochan, D. Z., et al. "Circadian disruption and breast cancer: An epigenetic link?" *Oncotarget* 6, no. 19 (2015): 16866.

12. Heikkinen, S., et al. "Does hair dye use increase the risk of breast cancer? A population-based case-control study of Finnish women." *PloS One* 10, no. 8 (2015): e0135190.

13. Rollison, D. E., et al. "Personal hair dye use and cancer: a systematic literature review and evaluation of exposure assessment in studies published since 1992." *Journal of Toxicology and Environmental Health, Part B* 9, no. 5 (2006): 413–39.

14. Takkouche, B., et al. "Personal use of hair dyes and risk of cancer: a meta-analysis." *JAMA* 293, no. 20 (2005): 2516–25.

15. Takkouche, B., et al. "Risk of cancer among hairdressers and related workers: a meta-analysis." *International Journal of Epidemiology* 38, no. 6 (2009): 1512–31.

16. "Progress cleaning the air and improving people's health," *United States Environmental Protection Agency*, accessed February 7, 2016, www.epa.gov/clean-air-act-overview/progress-cleaning-air-and-improving-peoples-health.

17. "Southern California air regulators fail to make decision on methane gas leak," *Fox News*, January 17, 2016, accessed February 7, 2016, www.foxnews.com/us/2016/01/17/southern-california-air-regulators-fail-to-make-decision-on-methane-gas-leak.html; "Methane from massive gas leak in Porter Ranch is boosting global warming: Experts," *Los Angeles Times*, January 24, 2016, accessed February 7, 2016, http://ktla.com/2016/01/24/methane-from-massive-gas-leak-in-porter-ranch-is-boosting-global-warming-experts/; "As California methane leak displaces thousands, will U.S. regulate natural gas sites nationwide," *Democracy Now*, January 14, 2016, accessed February 7, 2016, www.democracynow.org/2016/1/14/as_california_methane_leak_displaces_thousands.

18. "Erin Brockovich: California methane gas leak is worst U.S. environmental disaster since BP oil spill," *Democracy Now*, December 30, 2015, accessed February 7, 2016, www.democracynow.org/2015/12/30/erin_brockovich_california_methane_gas_leak.

19. Bell, M. L., et al. "Ozone and short-term mortality in 95 US urban communities, 1987–2000." *JAMA* 292, no. 19 (2004): 2372–78; Gryparis, A., et al. "Acute effects of ozone on mortality from the 'air pollution and health: a European approach' project." *American Journal of Respiratory and Critical Care Medicine* 170, no. 10 (2004): 1080–87; Bell, M. L., et al. "A meta-analysis of time-series studies of ozone and mortality with comparison to the national morbidity, mortality, and air pollution study." *Epidemiology* 16, no. 4 (2005): 436; Levy, J. I., et al. "Ozone exposure and mortality: an empiric bayes metaregression analysis." *Epidemiology* 16, no. 4 (2005): 458–68; Ito, K., et al. "Associations between ozone and daily mortality: analysis and meta-analysis." *Epidemiology* 16, no. 4 (2005): 446–57; Zanobetti, A., et al. "Mortality displacement in the association of ozone with mortality: an analysis of 48 cities in the United States." *American Journal of Respiratory and Critical Care Medicine*

177, no. 2 (2008): 184–89; Katsouyanni, K., et al. "Air pollution and health: a European and North American approach (APHENA)." *Research Report (Health Effects Institute)* 142 (2009): 5–90; Samoli, E., et al. "The temporal pattern of mortality responses to ambient ozone in the APHEA project." *Journal of Epidemiology and Community Health* 63, no. 12 (2009): 960–66.

20. Pope III, C. A., et al. "Health effects of fine particulate air pollution: lines that connect." *Journal of the Air and Waste Management Association* 56, no. 6 (2006): 709–42; Seaton, A., et al. "Particulate air pollution and acute health effects." *Lancet* 345, no. 8943 (1995): 176–78; Kim, J. J. "Ambient air pollution: health hazards to children." *Pediatrics* 114, no. 6 (2004): 1699–1707; "Health effects of ozone and particle pollution," *State of the Air*, accessed December 28, 2015, www.stateoftheair.org/2013/health-risks.

21. Calderón-Garcidueñas, L., et al. "Mexico City normal weight children exposed to high concentrations of ambient PM 2.5 show high blood leptin and endothelin-1, vitamin D deficiency, and food reward hormone dysregulation versus low pollution controls. Relevance for obesity and Alzheimer disease." *Environmental Research* 140 (2015): 579–92.

22. Morris, B. J., et al. "FOXO3: A major gene for human longevity—a mini-review." *Gerontology* (2015); Singh, R., et al. "Anti-inflammatory heat shock protein 70 genes are positively associated with human survival." *Current Pharmaceutical Design* 16, no. 7 (2010): 796.

23. Laukkanen, T., et al. "Association between sauna bathing and fatal cardiovascular and all-cause mortality events." *JAMA Internal Medicine* 175, no. 4 (2015): 542–48.

24. Kenttämies, A., et al. "Death in sauna." *Journal of Forensic Sciences* 53, no. 3 (2008): 724–29.

25. Scoon, G. S., et al. "Effect of post-exercise sauna bathing on the endurance performance of competitive male runners." *Journal of Science and Medicine in Sport* 10, no. 4 (2007): 259–62.

26. Stanley, J., et al. "Effect of sauna-based heat acclimation on plasma volume and heart rate variability." *European Journal of Applied Physiology* 115, no. 4 (2015): 785–94.

27. Krause, M., et al. "Heat shock proteins and heat therapy for type 2 diabetes: pros and cons." *Current Opinion in Clinical Nutrition and Metabolic Care* 18, no. 4 (2015): 374–80.

28. Kukkonen-Harjula, K., et al. "Haemodynamic and hormonal responses to heat exposure in a Finnish sauna bath." *European Journal of Applied Physiology and Occupational Physiology* 58, no. 5 (1989): 543–50.

29. Gryka, D., et al. "The effect of sauna bathing on lipid profile in young, physically active, male subjects." *International Journal of Occupational Medicine and Environmental Health* 27, no. 4 (2014): 608–18; Pilch, W., et al. "Changes in the lipid profile of blood serum in women taking sauna baths of various duration." *International Journal of Occupational Medicine and Environmental Health* 23, no. 2 (2010): 167–74; van der Wall, E. E. "Sauna bathing: a warm heart proves beneficial." *Netherlands Heart Journal* 23, no. 5 (2015): 247.

30. Tomiyama, C., et al. "The effect of repetitive mild hyperthermia on body temperature, the autonomic nervous system, and innate and adaptive immunity." *Biomedical Research* 36, no. 2 (2015): 135–42.

31. Hooper, L. V. "You AhR what you eat: linking diet and immunity." *Cell* 147, no. 3 (2011): 489–91.

32. Li, Y., et al. "Exogenous stimuli maintain intraepithelial lymphocytes via aryl hydrocarbon receptor activation." *Cell* 147, no. 3 (2011): 629–40.

33. Nishi, K., et al. "Immunostimulatory in vitro and in vivo effects of a water-soluble extract from kale." *Bioscience, Biotechnology, and Biochemistry* 75, no. 1 (2011): 40–46.

34. Haddad, E. H., et al. "Effect of a walnut meal on postprandial oxidative stress and antioxidants in healthy individuals." *Nutrition Journal* 13, no. 1 (2014): 1.

35. Cominetti, C., et al. "Associations between glutathione peroxidase-1 Pro198Leu polymorphism, selenium status, and DNA damage levels in obese women after consumption of Brazil nuts." *Nutrition* 27, no. 9 (2011): 891–96.

36. Song, J. M., et al. "Antiviral effect of catechins in green tea on influenza virus." *Antiviral Research* 68, no. 2 (2005): 66–74; Hsu, S. "Compounds derived from epigallocatechin-3-gallate (EGCG) as a novel approach to the prevention of viral infections." *Inflammation and Allergy-Drug Targets* 14, no. 1 (2015): 13–18; Rowe, C. A., et al. "Specific formulation of Camellia Sinensis prevents cold and flu symptoms and enhances γδ T cell function: a randomized, double-blind, placebo-controlled study." *Journal of the American College of Nutrition* 26, no. 5 (2007): 445–52.

37. Mak, J.C.W. "Potential role of green tea catechins in various disease therapies: progress and promise." *Clinical and Experimental Pharmacology and Physiology* 39, no. 3 (2012): 265–73; Apetz, N., et al. "Natural compounds and plant extracts as therapeutics against chronic inflammation in Alzheimer's disease-a translational perspective." *CNS and Neurological Disorders-Drug Targets (Formerly Current Drug Targets-CNS and Neurological Disorders)* 13, no. 7 (2014): 1175–91; Chen, G., et al. "Nutraceuticals and functional foods in the management of hyperlipidemia." *Critical Reviews in Food Science and Nutrition* 54, no. 9 (2014): 1180–1201; Johnson, R., et al. "Green tea and green tea catechin extracts: an overview of the clinical evidence." *Maturitas* 73, no. 4 (2012): 280–87. "Green tea extract for external anogenital warts." *Drug and Therapeutics Bulletin* 53, no. 10 (2015): 114–16; Gupta, A. K., et al. "Sinecatechins 10% ointment: A green tea extract for the treatment of external genital warts." *Pain* 46 (2015): 14–15; Tatti, S., et al. "Sinecatechins, a defined green tea extract, in the treatment of external anogenital warts: a randomized controlled trial." *Obstetrics and Gynecology* 111, no. 6 (2008): 1371–79.

38. Marinac, C. R., et al. "Frequency and circadian timing of eating may influence biomarkers of inflammation and insulin resistance associated with breast cancer risk." *PloS One* 10, no. 8 (2015): e0136240; Marinac, C. R., et al. "Prolonged nightly fasting and breast cancer risk: Findings from NHANES (2009–2010)."*Cancer Epidemiology Biomarkers and Prevention* 24, no. 5 (2015): 783–89.

39. Beitner, H. "Randomized, placebo-controlled, double blind study on the clinical efficacy of a cream containing 5% α-lipoic acid related to photoageing of facial skin." *British Journal of Dermatology* 149, no. 4 (2003): 841–49.

40. Moura, F. A., et al. "Lipoic acid: its antioxidant and anti-inflammatory role and clinical applications." *Current Topics in Medicinal Chemistry* 15, no. 5 (2015): 458–83; Patel, M. K., et al. "Can α-lipoic acid mitigate progression of aging-related decline caused by oxidative stress?" *Southern Medical Journal* 107, no. 12 (2014): 780–87.

41. Maczurek, A., et al. "Lipoic acid as an anti-inflammatory and neuroprotective treatment for Alzheimer's disease." *Advanced Drug Delivery Reviews* 60, no. 13 (2008): 1463–70.

42. Roberts, J. L., et al. "Emerging role of alpha-lipoic acid in the prevention and treatment of bone loss." *Nutrition Reviews* 73, no. 2 (2015): 116–25.

43. Kouzi, S. A., et al. "Natural supplements for improving insulin sensitivity and glucose uptake in skeletal muscle." *Frontiers in Bioscience* 7 (2014): 94–106; Lee, T., et al. "Nutritional supplements and their effect on glucose control." *Advances in Experimental Medicine and Biology* 771, (2012): 381–95.

44. Huerta, A. E., et al. "Effects of α-lipoic acid and eicosapentaenoic acid in overweight and obese women during weight loss." *Obesity* 23, no. 2 (2015): 313–21.

45. Carbonelli, M. G., et al. "α-Lipoic acid supplementation: a tool for obesity therapy?" *Current Pharmaceutical Design* 16, no. 7 (2010): 840–46.

46. Koh, E. H., et al. "Effects of alpha-lipoic acid on body weight in obese subjects." *American Journal of Medicine* 124, no. 1 (2011): 85-e1.

47. Natural Resources Defense Council, "What's on Tap?" last revised 2/6/2012, www.nrdc.org/water/drinking/uscities/pdf/chap04.pdf, accessed February 15, 2016. See also www.nrdc.org/health/safe-drinking-water.asp. You can also look up your drinking water on the Environmental Working Group website, www.ewg.org/tap-water/whats-in-yourwater.php, accessed February 15, 2016.

Chapter 10: Soothe—Week 6

1. Epel, E. S., et al. "Accelerated telomere shortening in response to life stress." *Proceedings of the National Academy of Sciences of the United States of America* 101, no. 49 (2004): 17312–15.

2. Bonini, L. "The Extended Mirror Neuron Network Anatomy, Origin, and Functions." *Neuroscientist* (2016): 1073858415626400; Caramazza, A., et al. "Embodied cognition and mirror neurons: a critical assessment." *Annual Review of Neuroscience* 37 (2014): 1–15; Cook, R., et al. "Mirror neurons: from origin to function." *Behavioral and Brain Sciences* 37, no. 2 (2014): 177–92.

3. Acharya, S., et al. "Mirror neurons: Enigma of the metaphysical modular brain." *Journal of Natural Science, Biology and Medicine* 3, no. 2 (2012): 118.
4. Yehuda, R., et al. "Holocaust exposure induced intergenerational effects on FKBP5 methylation." *Biological Psychiatry* (2015).
5. Yehuda, R., et al. "Gene expression patterns associated with posttraumatic stress disorder following exposure to the World Trade Center attacks." *Biological Psychiatry* 66, no. 7 (2009): 708–11.
6. Dias, B. G., et al. "Parental olfactory experience influences behavior and neural structure in subsequent generations." *Nature Neuroscience* 17, no. 1 (2014): 89–96.
7. McEwen, B. S., et al. "Protective and damaging effects of stress mediators." *New England Journal of Medicine* 338, no. 3 (1998): 171–79; Jiang, W., et al. "Mental stress–induced myocardial ischemia and cardiac events." *JAMA* 275, no. 21 (1996): 1651–56; Deanfield, J. E., et al. "Silent myocardial ischaemia due to mental stress." *Lancet* 324, no. 8410 (1984): 1001–5; Rozanski, A., et al. "Mental stress and the induction of silent myocardial ischemia in patients with coronary artery disease." *New England Journal of Medicine* 318, no. 16 (1988): 1005–12; Nabi, H., et al. "Increased risk of coronary heart disease among individuals reporting adverse impact of stress on their health: the Whitehall II prospective cohort study." *European Heart Journal* (2013): eht216; Arnold, S. V., et al. "Perceived stress in myocardial infarction: long-term mortality and health status outcomes." *Journal of the American College of Cardiology* 60, no. 18 (2012): 1756–63; Richardson, S., et al. "Meta-analysis of perceived stress and its association with incident coronary heart disease." *American Journal of Cardiology* 110, no. 12 (2012): 1711–16; Deedwania, P. C. "Editorial comment: Mental stress, pain perception and risk of silent ischemia." *Journal of the American College of Cardiology* 25, no. 7 (1995): 1504–6; Steptoe, A., et al. "Stress and cardiovascular disease." *Nature Reviews Cardiology* 9, no. 6 (2012): 360–70; Alevizos, M., et al. "Stress triggers coronary mast cells leading to cardiac events." *Annals of Allergy, Asthma and Immunology* 112, no. 4 (2014): 309–16; Chida, Y., et al. "A bidirectional relationship between psychosocial factors and atopic disorders: a systematic review and meta-analysis." *Psychosomatic Medicine* 70, no. 1 (2008): 102–16; Theoharides, T. C., et al. "Critical role of mast cells in inflammatory diseases and the effect of acute stress." *Journal of Neuroimmunology* 146, no. 1 (2004): 1–12; Wright, R., et al. "The impact of stress on the development and expression of atopy." *Current Opinion in Allergy and Clinical Immunology* 5, no. 1 (2005): 23–29; Slattery, M. J. "Psychiatric comorbidity associated with atopic disorders in children and adolescents." *Immunology and Allergy Clinics of North America* 25, no. 2 (2005): 407–20; Seiffert, K., et al. "Psychophysiological reactivity under mental stress in atopic dermatitis." *Dermatology* 210, no. 4 (2005): 286293; Chen, E., et al. "Stress and inflammation in exacerbations of asthma." *Brain, Behavior, and Immunity* 21, no. 8 (2007): 993–99; Theoharides, T. C., et al. "Contribution of stress to asthma worsening through mast cell activation." *Annals of Allergy, Asthma and Immunology* 109, no. 1 (2012): 14–19.
8. Judge, T. A., et al. "Genetic influences on core self-evaluations, job satisfaction, and work stress: A behavioral genetics mediated model." *Organizational Behavior and Human Decision Processes* 117, no. 1 (2012): 208–20.
9. Ising, M., et al. "Genetics of stress response and stress-related disorders." *Dialogues in Clinical Neuroscience* 8, no. 4 (2006): 433.
10. Slominski, A. T., et al. "Key role of CRF in the skin stress response system." *Endocrine Reviews* 34, no. 6 (2013): 827–84.
11. Ising, M., et al. "Polymorphisms in the FKBP5 gene region modulate recovery from psychosocial stress in healthy controls." *European Journal of Neuroscience* 28, no. 2 (2008): 389–98.
12. Zuhl, M., et al. "Exercise regulation of intestinal tight junction proteins." *British Journal of Sports Medicine* (2012): bjsports-2012.
13. Piacentini, M. F., et al. "Stress related changes during a half marathon in master endurance athletes." *Journal of Sports Medicine and Physical Fitness* 55, no. 4 (2015): 329–36; Brisswalter, J., et al. "Neuromuscular factors associated with decline in long-distance running performance in master athletes." *Sports Medicine* 43, no. 1 (2013): 51–63; Lac, G., et al.

"Changes in cortisol and testosterone levels and T/C ratio during an endurance competition and recovery." *Journal of Sports Medicine and Physical Fitness* 40, no. 2 (2000): 139.

14. Lamprecht, M., et al. "Exercise, intestinal barrier dysfunction and probiotic supplementation." *Medicine and Sport Science* (2012): 47–56; Lamprecht, M., et al. "Probiotic supplementation affects markers of intestinal barrier, oxidation, and inflammation in trained men; a randomized, double-blinded, placebo-controlled trial." *Journal of the International Society of Sports Nutrition* 9, no. 1 (2012): 45.

15. Delarue, J., et al. "Fish oil prevents the adrenal activation elicited by mental stress in healthy men." *Diabetes and Metabolism* 29, no. 3 (2003): 289–95.

16. Peters, E. M., et al. "Vitamin C supplementation attenuates the increases in circulating cortisol, adrenaline and anti-inflammatory polypeptides following ultramarathon running." *International Journal of Sports Medicine* 22, no. 7 (2001): 537–43; Peters, E. M., et al. "Attenuation of increase in circulating cortisol and enhancement of the acute phase protein response in vitamin C-supplemented ultramarathoners." *International Journal of Sports Medicine* 22, no. 2 (2001): 120–26.

17. Bryant, E. F. *The Yoga Sutras of Patanjali: A New Edition, Translation, and Commentary* (New York: North Point Press, 2009).

18. Luders, E., et al. "The underlying anatomical correlates of long-term meditation: larger hippocampal and frontal volumes of gray matter." *Neuroimage* 45, no. 3 (2009): 672–78.

19. Slagter, H. A., et al. "Theta phase synchrony and conscious target perception: impact of intensive mental training." *Journal of Cognitive Neuroscience* 21, no. 8 (2009): 1536–49.

20. Albert, E. *After Birth* (Boston: Houghton Mifflin Harcourt, 2015), 112.

21. Porges, S. W. "The polyvagal theory: new insights into adaptive reactions of the autonomic nervous system." *Cleveland Clinic Journal of Medicine* 76, Suppl. 2 (2009): S86; Danner, D. D., et al. "Positive emotions in early life and longevity: findings from the nun study." *Journal of Personality and Social Psychology* 80, no. 5 (2001): 804.

22. Mäkinen, T. M., et al. "Autonomic nervous function during whole-body cold exposure before and after cold acclimation." *Aviation, Space, and Environmental Medicine* 79, no. 9 (2008): 875–82.

23. Lu, W. A., et al. "Foot reflexology can increase vagal modulation, decrease sympathetic modulation, and lower blood pressure in healthy subjects and patients with coronary artery disease." *Alternative Therapies in Health and Medicine* 17, no. 4 (2011): 8–14.

24. Yang, J. L., et al. "Comparison of effect of 5 recumbent positions on autonomic nervous modulation in patients with coronary artery disease." *Circulation Journal* 72, no. 6 (2008): 902–08.

25. Vickhoff, B., et al. "Music structure determines heart rate variability of singers." *Frontiers in Psychology* 4 (2013).

26. Richards, D., et al. "Stimulation of auricular acupuncture points in weight loss." *Australian Family Physician* 27 (1998): S73–7; Yang, S. B., et al. "Efficacy comparison of different points combination in the treatment of menopausal insomnia: a randomized controlled trial." *Zhongguo Zhen Jiu [Chinese Acupuncture and Moxibustion]* 34, no. 1 (2014): 3–8; da Silva, M.A.H., et al. "Neuroanatomic and clinical correspondences: acupuncture and vagus nerve stimulation." *Journal of Alternative and Complementary Medicine* 20, no. 4 (2014): 233–40; He, W., et al. "Auricular acupuncture and vagal regulation." *Evidence-Based Complementary and Alternative Medicine* (2012).

27. Girsberger, W., et al. "Heart rate variability and the influence of craniosacral therapy on autonomous nervous system regulation in persons with subjective discomforts: a pilot study." *Journal of Integrative Medicine* 12, no. 3 (2014): 156–61.

28. Kabat-Zinn, J. "Mindfulness-based interventions in context: past, present, and future." *Clinical Psychology: Science and Practice* 10, no. 2 (2003): 144–56.

29. Davidson, R. J. "Affective style, psychopathology, and resilience: brain mechanisms and plasticity." *American Psychologist* 55, no. 11 (2000): 1196; Davidson, R. J., et al. "Alterations in brain and immune function produced by mindfulness meditation." *Psychosomatic Medicine* 65, no. 4 (2003): 564–70.

30. Fredrickson, B. L., et al. "Open hearts build lives: positive emotions, induced through loving-kindness meditation, build consequential personal resources." *Journal of Personality and*

Social Psychology 95, no. 5 (2008): 1045; Kalyani, B. G., et al. "Neurohemodynamic correlates of 'OM' chanting: a pilot functional magnetic resonance imaging study." *International Journal of Yoga* 4, no. 1 (2011): 3; Mason, H., et al. "Cardiovascular and respiratory effect of yogic slow breathing in the yoga beginner: what is the best approach?" *Evidence-Based Complementary and Alternative Medicine* (2013); Khalsa, D. S., et al. "Cerebral blood flow changes during chanting meditation." *Nuclear Medicine Communications* 30, no. 12 (2009): 956–61; Chang, R. Y., et al. "The effect of t'ai chi exercise on autonomic nervous function of patients with coronary artery disease." *Journal of Alternative and Complementary Medicine* 14, no. 9 (2008): 1107–13.

31. Leung, M.K., et al. "Increased gray matter volume in the right angular and posterior parahippocampal gyri in loving-kindness meditators." *Social Cognitive and Affective Neuroscience* (2012): nss076; Kang, D. H., et al. "The effect of meditation on brain structure: cortical thickness mapping and diffusion tensor imaging." *Social Cognitive and Affective Neuroscience* 8, no. 1 (2013): 27–33; Lazar, S. W., et al. "Meditation experience is associated with increased cortical thickness." *Neuroreport* 16, no. 17 (2005): 1893.

32. Macartney, M. J., et al. "Intrinsic heart rate recovery after dynamic exercise is improved with an increased omega-3 index in healthy males." *British Journal of Nutrition* 112, no. 12 (2014): 1984–92; Ninio, D. M., et al. "Docosahexaenoic acid-rich fish oil improves heart rate variability and heart rate responses to exercise in overweight adults." *British Journal of Nutrition* 100, no. 05 (2008): 1097–1103; Sjoberg, N. J., et al. "Dose-dependent increases in heart rate variability and arterial compliance in overweight and obese adults with DHA-rich fish oil supplementation." *British Journal of Nutrition* 103, no. 2 (2010): 243–48; Xin, W., et al. "Short-term effects of fish-oil supplementation on heart rate variability in humans: a meta-analysis of randomized controlled trials." *American Journal of Clinical Nutrition* 97, no. 5 (2013): 926–35; Noreen, E. E., et al. "Effects of supplemental fish oil on resting metabolic rate, body composition, and salivary cortisol in healthy adults." *Journal of the International Society of Sports Nutrition* 7, no. 31 (2010); Delarue et al. "Fish oil prevents the adrenal activation."

33. Keating, T. *Intimacy with God: An Introduction to Centering Prayer* (Spring Valley, NY: Crossroad, 2009); Bourgeault, C. *Centering Prayer and Inner Awakening* (Lanham, MD: Rowman and Littlefield, 2004).

34. Hölzel, B. K., et al. "Stress reduction correlates with structural changes in the amygdala." *Social Cognitive and Affective Neuroscience* (2009): nsp034.

Chapter 11: Think—Week 7

1. Stone, A. A., et al. "A snapshot of the age distribution of psychological well-being in the United States." *Proceedings of the National Academy of Sciences* 107, no. 22 (2010): 9985–90.

2. O'Donohue, J. *Beauty: The Invisible Embrace* (New York: HarperCollins, 2004).

3. Brach, T. "Working with Difficulties: The Blessings of RAIN," *TaraBrach.com*, accessed March 24, 2016. www.tarabrach.com/articles-interviews/rain-workingwithdifficulties.

4. Tippett, K. "John O'Donohue—the inner landscape of beauty," *On Being with Krista Tippett*, August 6, 2015, accessed February 10, 2016. www.onbeing.org/program/john-o-donohue-the-inner-landscape-beauty/203.

5. Dadhania, V. P., et al. "Nutraceuticals against neurodegeneration: A mechanistic insight." *Current Neuropharmacology* (2016).

6. Amor, S., et al. "Inflammation in neurodegenerative diseases." *Immunology* 129, no. 2 (2010): 154–69.

7. Fischer, R., et al. "Interrelation of oxidative stress and inflammation in neurodegenerative disease: role of TNF." *Oxidative Medicine and Cellular Longevity* (2015); Hooshmand, B., et al. "Homocysteine and holotranscobalamin and the risk of Alzheimer disease a longitudinal study." *Neurology* 75, no. 16 (2010): 1408–14; Laurin, D., et al. "Midlife C-reactive protein and risk of cognitive decline: a 31-year follow-up." *Neurobiology of Aging* 30, no. 11 (2009): 1724–27; Komulainen, P., et al. "Serum high sensitivity C-reactive protein and cognitive function in elderly women." *Age and Ageing* 36, no. 4 (2007): 443–48; Alcolea, D., et al.

"Amyloid precursor protein metabolism and inflammation markers in preclinical Alzheimer disease." *Neurology* 85, no. 7 (2015): 626–33.

8. Duchen, M. R. "Mitochondria and calcium: from cell signalling to cell death." *Journal of Physiology* 529, no. 1 (2000): 57–68; Marambaud, P., et al "Calcium signaling in neurodegeneration." *Molecular Neurodegeneration* 4, no. 20 (2009): 6–5; Bezprozvanny, I. B. "Calcium signaling and neurodegeneration." *Acta Naturae* 2, no. 1 (2010): 72.

9. Toescu, E. C., et al. "The importance of being subtle: small changes in calcium homeostasis control cognitive decline in normal aging." *Aging Cell* 6, no. 3 (2007): 267–73.

10. Sun, A. Y., et al. "Oxidative stress and neurodegenerative disorders." *Journal of Biomedical Science* 5, no. 6 (1998): 401–14; López-Armada, M. J., et al. "Mitochondrial dysfunction and the inflammatory response." *Mitochondrion* 13, no. 2 (2013): 106–18.

11. Lane, R. K., et al. "The role of mitochondrial dysfunction in age-related diseases." *Biochimica et Biophysica Acta (BBA)-Bioenergetics* 1847, no. 11 (2015): 1387–1400.

12. Lin, M. T., et al. "Mitochondrial dysfunction and oxidative stress in neurodegenerative diseases." *Nature* 443, no. 7113 (2006): 787–95; Petrozzi, L., et al. "Mitochondria and neurodegeneration." *Bioscience Reports* 27 (2007): 87–104; Kaminsky, Y. G., et al. "Critical analysis of Alzheimer's amyloid-beta toxicity to mitochondria." *Frontiers in Bioscience* 20 (2015): 173–97; Lionaki, E., et al. "Mitochondria, autophagy and age-associated neurodegenerative diseases: New insights into a complex interplay."*Biochimica et Biophysica Acta (BBA)-Bioenergetics* (2015).

13. Alzheimer's Association. "2015 Alzheimer's disease facts and figures." *Alzheimer's and Dementia: Journal of the Alzheimer's Association* 11, no. 3 (2015): 332.

14. "What we know today about Alzheimer's Disease," *Alzheimer's Association*, accessed January 9, 2016, www.alz.org/research/science/alzheimers_disease_causes.asp.

15. Kuro-o, M. "Klotho as a regulator of oxidative stress and senescence." *Biological Chemistry* 389, no. 3 (2008): 233–41; Mitobe, M., et al. "Oxidative stress decreases klotho expression in a mouse kidney cell line." *Nephron Experimental Nephrology* 101, no. 2 (2005): e67–e74; Yamamoto, M., et al. "Regulation of oxidative stress by the anti-aging hormone klotho." *Journal of Biological Chemistry* 280, no. 45 (2005): 38029–34; Troyano-Suárez, N., et al. "Glucose oxidase induces cellular senescence in immortal renal cells through ILK by downregulating klotho gene expression." *Oxidative Medicine and Cellular Longevity* (2015); Kim, J. H., et al. "Biological role of anti-aging protein klotho." *Journal of Lifestyle Medicine* 5, no. 1 (2015): 1.

16. Kuro-o, M., et al. "Mutation of the mouse klotho gene leads to a syndrome resembling ageing." *Nature* 390, no. 6655 (1997): 45–51.

17. "2015 Alzheimer's disease facts and figures," *Alzheimer's Association*.

18. Mayeux, R., et al. "Epidemiology of Alzheimer disease." *Cold Spring Harbor Perspectives in Medicine* 2, no. 8 (2012): a006239.

19. Blennow K., et al. "Alzheimer's disease." *Lancet*, 368, no. 9533 (2006): 387–403.

20. Loy, C. T., et al. "Genetics of dementia." *Lancet* 383, no. 9919 (2014): 828–40; Holtzman, D. M., et al. "Apolipoprotein E and apolipoprotein E receptors: normal biology and roles in Alzheimer disease." *Cold Spring Harbor Perspectives in Medicine* 2, no. 3 (2012): a006312.

21. Theendakara, V., et al. "Neuroprotective sirtuin ratio reversed by ApoE4." *Proceedings of the National Academy of Sciences* 110, no. 45 (2013): 18303–8; Tramutola, A., et al. "Alteration of mTOR signaling occurs early in the progression of Alzheimer disease (AD): analysis of brain from subjects with pre-clinical AD, amnestic mild cognitive impairment and late-stage AD." *Journal of Neurochemistry* 133, no. 5 (2015): 739–49.

22. Bredesen, D. E. "Reversal of cognitive decline: a novel therapeutic program." *Aging* 6, no. 9 (2014): 707; Bredesen, D. E., et al. "Reversal of cognitive decline in Alzheimer's disease." *Aging* 8, no. 6 (2016): 1250.

23. Bredesen, D. E. "Metabolic profiling distinguishes three subtypes of Alzheimer's disease." *Aging* 7, no. 8 (2015): 595; Bredesen, D. E. "Inhalational Alzheimer's disease: an unrecognized—and treatable—epidemic." *Aging* 8, no. 2 (2016): 304.

24. Gatz, M., et al. "Role of genes and environments for explaining Alzheimer disease." *Archives of General Psychiatry* 63, no. 2 (2006): 168–74; Kamboh, M. I., et al. "Genome-wide association study of Alzheimer's disease." *Translational Psychiatry* 2, no. 5 (2012): e117.

25. Sleegers, K., et al. "The pursuit of susceptibility genes for Alzheimer's disease: progress and prospects." *Trends in Genetics* 26, no. 2 (2010): 84–93.

26. Fraga, M. F., et al. "Epigenetic differences arise during the lifetime of monozygotic twins." *Proceedings of the National Academy of Sciences of the United States of America* 102, no. 30 (2005): 10604–9; Heyn, H., et al. "Distinct DNA methylomes of newborns and centenarians." *Proceedings of the National Academy of Sciences* 109, no. 26 (2012): 10522–27.

27. Bredesen, D. E., et al. "Next generation therapeutics for Alzheimer's disease." *EMBO Molecular Medicine* 5, no. 6 (2013): 795–98; Yaffe, K., et al. "Estrogen use, APOE, and cognitive decline evidence of gene–environment interaction." *Neurology* 54, no. 10 (2000): 1949–54.

28. "2005 adult sleep habits and styles," *National Sleep Foundation*, accessed February 1, 2016, https://sleepfoundation.org/sleep-polls-data/sleep-in-america-poll/2005-adult-sleep-habits-and-styles.

29. Smith, G. E., et al. "A cognitive training program based on principles of brain plasticity: Results from the Improvement in Memory with Plasticity-Based Adaptive Cognitive Training (IMPACT) Study." *Journal of the American Geriatrics Society* 57, no. 4 (2009): 594–603.

30. Hatch, S. L., et al. "The continuing benefits of education: adult education and midlife cognitive ability in the British 1946 birth cohort." *Journals of Gerontology Series B: Psychological Sciences and Social Sciences* 62, no. 6 (2007): S404–14.

31. Woods, B., et al. "Cognitive stimulation to improve cognitive functioning in people with dementia." *Cochrane Database of Systematic Reviews* 2 (2012).

32. Schmidt, S. R. "Effects of humor on sentence memory." *Journal of Experimental Psychology: Learning, Memory, and Cognition* 20, no. 4 (1994): 953.

33. Ybarra, O., et al. "Mental exercising through simple socializing: Social interaction promotes general cognitive functioning." *Personality and Social Psychology Bulletin* 34, no. 2 (2008): 248–59.

34. Bassuk, S. S., et al. "Social disengagement and incident cognitive decline in community-dwelling elderly persons." *Annals of Internal Medicine* 131, no. 3 (1999): 165–73.

35. Savignac, H. M., et al. "Prebiotic feeding elevates central brain derived neurotrophic factor, N-methyl-D-aspartate receptor subunits and D-serine." *Neurochemistry International* 63, no. 8 (2013): 756–64.

36. Beck, J. "Your gut bacteria want you to eat a cupcake," *Atlantic* August 19, 2014, accessed February 1, 2016, www.theatlantic.com/health/archive/2014/08/your-gut-bacteria-want-you-to-eat-a-cupcake/378702.

37. Kivipelto, M., et al. "Alzheimer disease: To what extent can Alzheimer disease be prevented?" *Nature Reviews Neurology* 10, no. 10 (2014): 552–53.

38. Ibid.; Kawas, C., et al. "Age-specific incidence rates of Alzheimer's disease: The Baltimore Longitudinal Study of Aging." *Neurology* 54, no. 11 (2000): 2072–77.

39. Smith, J., et al. "Physical activity reduces hippocampal atrophy in elders at genetic risk for Alzheimer's disease." *Frontiers in Aging Neuroscience* 6 (2014).

40. Sui, X., et al. "A prospective study of cardiorespiratory fitness and risk of type 2 diabetes in women." *Diabetes Care* 31, no. 3 (2008): 550–55; Blair, S. N., et al. "Physical fitness and all-cause mortality: a prospective study of healthy men and women." *JAMA* 262, no. 17 (1989): 2395–2401; Hooker, S. P., et al. "Cardiorespiratory fitness as a predictor of fatal and nonfatal stroke in asymptomatic women and men." *Stroke* 39, no. 11 (2008): 2950–57; Wei, M., et al. "The association between cardiorespiratory fitness and impaired fasting glucose and type 2 diabetes mellitus in men." *Annals of Internal Medicine* 130, no. 2 (1999): 89–96.

41. DeFina, L. F., et al. "The association between midlife cardiorespiratory fitness levels and later-life dementia: a cohort study." *Annals of Internal Medicine* 158, no. 3 (2013): 162–68.

42. Smith, G. E., et al. "A cognitive training program based on principles of brain plasticity: Results from the Improvement in Memory with Plasticity-based Adaptive Cognitive Training (IMPACT) Study." *Journal of the American Geriatrics Society* 57, no. 4 (2009): 594–603.

43. Cotman, C. W., et al. "Exercise builds brain health: key roles of growth factor cascades and inflammation." *Trends in Neurosciences* 30, no. 9 (2007): 464–72; Aguiar, P., et

al. "Rivastigmine transdermal patch and physical exercises for Alzheimer's disease: a randomized clinical trial." *Current Alzheimer Research* 11, no. 6 (2014): 532–37.

44. Etnier, J. L., et al. "A meta-regression to examine the relationship between aerobic fitness and cognitive performance." *Brain Research Reviews* 52, no. 1 (2006): 119–30; Angevaren, M., et al. "Physical activity and enhanced fitness to improve cognitive function in older people without known cognitive impairment." *Cochrane Database System Review* 3, no. 3 (2008); Erickson, K. I., et al. "Exercise training increases size of hippocampus and improves memory." *Proceedings of the National Academy of Sciences* 108, no. 7 (2011): 3017–22; Woodard, J. L., et al. "Lifestyle and genetic contributions to cognitive decline and hippocampal integrity in healthy aging." *Current Alzheimer Research* 9, no. 4 (2012): 436.

45. Eny, K. M., et al. "Genetic variant in the glucose transporter type 2 is associated with higher intakes of sugars in two distinct populations." *Physiological Genomics* 33, no. 3 (2008): 355–60.

46. Matthews, G., et al. *Personality Traits* (Cambridge: Cambridge University Press, 2003).

47. Harvey, C. J., et al. "Who is predisposed to insomnia: a review of familial aggregation, stress-reactivity, personality and coping style." *Sleep Medicine Reviews* 18, no. 3 (2014): 237–47; Smith, T. W., et al. "Hostility, anger, aggressiveness, and coronary heart disease: An interpersonal perspective on personality, emotion, and health." *Journal of Personality* 72, no. 6 (2004): 1217–70; Suarez, E. C., et al. "The relation of aggression, hostility, and anger to lipopolysaccharide-stimulated tumor necrosis factor (TNF)-α by blood monocytes from normal men." *Brain, Behavior, and Immunity* 16, no. 6 (2002): 675–84; Jylhä, P., et al. "The relationship of neuroticism and extraversion to symptoms of anxiety and depression in the general population." *Depression and Anxiety* 23, no. 5 (2006): 281–89.

48. Akram, U., et al. "Anxiety mediates the relationship between perfectionism and insomnia symptoms: A longitudinal study." *PloS One* 10, no. 10 (2015): e0138865; Vincent, N. K., et al. "Perfectionism and chronic insomnia." *Journal of Psychosomatic Research* 49, no. 5 (2000): 349–54; de Azevedo, M. H. P., et al. "Perfectionism and sleep disturbance." *World Journal of Biological Psychiatry* 10, no. 3 (2009): 225–33; Azevedo, M. H., et al. "Longitudinal study on perfectionism and sleep disturbance." *World Journal of Biological Psychiatry* 11, no. 2 (2010): 476–85; van de Laar, M., et al. "The role of personality traits in insomnia." *Sleep Medicine Reviews* 14, no. 1 (2010): 61–68; Schramm, E., et al. "Mental comorbidity of chronic insomnia in general practice attenders using DSM-III-R." *Acta Psychiatrica Scandinavica* 91, no. 1 (1995): 10–17.

49. www.ncbi.nlm.nih.gov/pubmed/8893314; www.researchgate.net/profile/Laura_Richman/publication/7701075_Positive_emotion_and_health_Going_beyond_the_negative/links/0046351a8c8c404c29000000.pdf.

50. Gold, S. M., et al. "Basal serum levels and reactivity of nerve growth factor and brain-derived neurotrophic factor to standardized acute exercise in multiple sclerosis and controls." *Journal of Neuroimmunology* 138, no. 1 (2003): 99–105; Rojas Vega, S., et al. "Acute BDNF and cortisol response to low intensity exercise and following ramp incremental exercise to exhaustion in humans." *Brain Research* 1121, no. 1 (2006): 59–65; Ferris, L. T., et al. "The effect of acute exercise on serum brain-derived neurotrophic factor levels and cognitive function." *Medicine and Science in Sports and Exercise* 39, no. 4 (2007): 728–34; Tang, S. W., et al. "Influence of exercise on serum brain-derived neurotrophic factor concentrations in healthy human subjects." *Neuroscience Letters* 431, no. 1 (2008): 62–65; Gustafsson, G., et al. "The acute response of plasma brain-derived neurotrophic factor as a result of exercise in major depressive disorder." *Psychiatry Research* 169, no. 3 (2009): 244–48; Schmolesky, M. T., et al. "The effects of aerobic exercise intensity and duration on levels of brain-derived neurotrophic factor in healthy men." *Journal of Sports Science and Medicine* 12, no. 3 (2013): 502; de Melo Coelho, F. G., et al. "Acute aerobic exercise increases brain-derived neurotrophic factor levels in elderly with Alzheimer's disease." *Journal of Alzheimer's Disease* 39, no. 2 (2014): 401; Saucedo-Marquez, C. M., et al. "High intensity interval training evokes larger serum BDNF levels compared to intense continuous exercise." *Journal of Applied Physiology* (2015): jap-00126.

51. Fernando, W., et al. "The role of dietary coconut for the prevention and treatment of Alzheimer's disease: potential mechanisms of action." *British Journal of Nutrition* (2015):

1–14; Rebello, C. J., et al. "Pilot feasibility and safety study examining the effect of medium chain triglyceride supplementation in subjects with mild cognitive impairment: A randomized controlled trial." *BBA Clinical* 3 (2015): 123–25.

52. Reger, M. A., et al. "Effects of beta-hydroxybutyrate on cognition in memory-impaired adults." *Neurobiology of Aging* 25, no. 3 (2004): 311–14.

53. Zong, G., et al. "Frequent consumption of meals prepared at home and risk of Type 2 diabetes among American men and women." *Circulation* 132, Suppl 3 (2015): A17285.

54. Haze, S., et al. "Effects of fragrance inhalation on sympathetic activity in normal adults." *Japanese Journal of Pharmacology* 90, no. 3 (2002): 247253.

55. Lee, I. S., et al. "Effects of lavender aromatherapy on insomnia and depression in women college students." *Taehan Kanho Hakhoe Chi* 36, no. 1 (2006): 136–43; Shiina, Y., et al. "Relaxation effects of lavender aromatherapy improve coronary flow velocity reserve in healthy men evaluated by transthoracic Doppler echocardiography." *International Journal of Cardiology* 129, no. 2 (2008): 193–97; Lytle, J., et al. "Effect of lavender aromatherapy on vital signs and perceived quality of sleep in the intermediate care unit: a pilot study." *American Journal of Critical Care* 23, no. 1 (2014): 24–29; Lillehei, A. S., et al. "A systematic review of the effect of inhaled essential oils on sleep." *Journal of Alternative and Complementary Medicine* 20, no. 6 (2014): 441–51.

56. Igarashi, M., et al. "Effect of olfactory stimulation by fresh rose flowers on autonomic nervous activity." *Journal of Alternative and Complementary Medicine* 20, no. 9 (2014): 727–31.

57. World Health Organization, *Nutritional Anemias: Report of a Scientific Group,* World Health Organization Technical Report Series 405 (1968).

58. Tucker, K. L., et al. "Plasma vitamin B-12 concentrations relate to intake source in the Framingham Offspring Study," *American Journal of Clinical Nutrition* 71 (2000): 514–22.

59. Pawlak, R., et al. "How prevalent is vitamin B(12) deficiency among vegetarians?" *Nutritional Review* 71, no. 2 (February 2013): 110–17.

60. Tucker, "Plasma vitamin B-12 concentrations," and Tucker, K. L., et al. "Low plasma vitamin B12 is associated with lower BMD: The Framingham Osteoporosis Study," *Journal of Bone and Mineral Research* 20, no. 1 (January 2005): 152–58.

61. Knott, V., et al. "Neurocognitive effects of acute choline supplementation in low, medium and high performer healthy volunteers." *Pharmacology Biochemistry and Behavior* 131 (2015): 119–29.

62. Overgaard, K. "The effects of citicoline on acute ischemic stroke: A review." *Journal of Stroke and Cerebrovascular Diseases* 23, no. 7 (2014): 1764–69.

63. "Horizon: How video games can change your brain," *BBC* September 16, 2015, accessed January 21, 2016, www.bbc.com/news/technology-34255492; Churchland, P. "Videogames for seniors boost brainpower," *Wall Street Journal,* September 20, 2015, accessed January 21, 2016, www.wsj.com/articles/videogames-for-seniors-boost-brainpower-1443623158.

Chapter 12: Integrate

1. Painter, R. C., et al. "Transgenerational effects of prenatal exposure to the Dutch famine on neonatal adiposity and health in later life." *BJOG: An International Journal of Obstetrics and Gynaecology* 115, no. 10 (2008): 1243–49.

2. de Rooij, S. R., et al. "Prenatal undernutrition and cognitive function in late adulthood." *Proceedings of the National Academy of Sciences* 107, no. 39 (2010): 16881–86.

3. www.ncbi.nlm.nih.gov/pubmed/27146370 Bleker 2016.

4. Tobi, E. W., et al. "Early gestation as the critical time-window for changes in the prenatal environment to affect the adult human blood methylome." *International Journal of Epidemiology* (2015): dyv043.

5. Veenendaal, M.V.E., et al. "Transgenerational effects of prenatal exposure to the 1944–45 Dutch famine." *BJOG: An International Journal of Obstetrics and Gynaecology* 120, no. 5 (2013): 548–54.

6. Ekamper, P., et al. "Prenatal famine exposure and adult mortality from cancer, cardiovascular disease, and other causes through age 63 years." *American Journal of Epidemiology* 181, no. 4 (2015): 271–79.

7. de Rooij, S. R., et al. "Prenatal undernutrition and leukocyte telomere length in late adulthood: the Dutch famine birth cohort study." *American Journal of Clinical Nutrition* 102, no. 3 (2015): 655–60.

8. Halstead, R. "Study: Marin County men have the highest life expectancy in the nation, women rank no. 2," *Marin Independent Journal*, April 19, 2012, accessed January 31, 2016, www.marinij.com/article/ZZ/20120419/NEWS/120418529.

9. Levy, B. R., et al. "Association between positive age stereotypes and recovery from disability in older persons." *JAMA* 308, no. 19 (2012): 1972–73; Levy, B. R., et al. "Longitudinal benefit of positive self-perceptions of aging on functional health." *Journals of Gerontology Series B: Psychological Sciences and Social Sciences* 57, no. 5 (2002): P409–17; Sargent-Cox, K. A., et al. "The relationship between change in self-perceptions of aging and physical functioning in older adults." *Psychology and Aging* 27, no. 3 (2012): 750.

10. Levy, B.R., et al. "Subliminal strengthening improving older individuals' physical function over time with an implicit-age-stereotype intervention." *Psychological Science* 25, no. 12 (2014): 2127–35.

11. Levy, B. R., et al. "The stereotype-matching effect: greater influence on functioning when age stereotypes correspond to outcomes." *Psychology and Aging* 24, no. 1 (2009): 230.

12. Murphy, S. T., et al. "Additivity of nonconscious affect: combined effects of priming and exposure." *Journal of Personality and Social Psychology* 69, no. 4 (1995): 589.

13. Schneier, M. "Fashion's gaze turned to Joan Didion in 2015," *New York Times*, December 18, 2015, accessed December 26, 2015, www.nytimes.com/2015/12/20/fashion/joan-didion-celine-fashion-gaze.html.

14. Cooke, R. "Joan Didion as the new face of Celine? That's so smart," *Guardian* January 11, 2015, accessed December 26, 2015, www.theguardian.com/commentisfree/2015/jan/11/joan-didion-new-face-celine-smart.

15. Levy, B. R., et al. "Longevity increased by positive self-perceptions of aging." *Journal of Personality and Social Psychology* 83, no. 2 (2002): 261.

Top Seven Genes: What to Do

1. Read more at www.snpedia.com/index.php/Orientation.

INDEX

ABOUT THE AUTHOR

DR. SARA GOTTFRIED is a world-renowned leading health expert. Her first two books, *The Hormone Cure* (Scribner, March 2013) and *The Hormone Reset Diet* (HarperOne, 2015), were *New York Times* bestsellers and have sold hundreds of thousands of copies. After graduating from the physician-scientist training program at Harvard Medical School and from MIT, Dr. Gottfried completed her residency at the University of California at San Francisco. She lives in San Francisco.

ALSO BY SARA GOTTFRIED, M.D.